SpringerBriefs in Economics

SpringerBriefs present concise summaries of cutting-edge research and practical applications across a wide spectrum of fields. Featuring compact volumes of 50 to 125 pages, the series covers a range of content from professional to academic. Typical topics might include:

- A timely report of state-of-the art analytical techniques
- A bridge between new research results, as published in journal articles, and a contextual literature review
- A snapshot of a hot or emerging topic
- An in-depth case study or clinical example
- A presentation of core concepts that students must understand in order to make independent contributions

SpringerBriefs in Economics showcase emerging theory, empirical research, and practical application in microeconomics, macroeconomics, economic policy, public finance, econometrics, regional science, and related fields, from a global author community.

Briefs are characterized by fast, global electronic dissemination, standard publishing contracts, standardized manuscript preparation and formatting guidelines, and expedited production schedules.

More information about this series at http://www.springer.com/series/8876

Philipp Aerni

Global Business in Local Culture

The Impact of Embedded Multinational Enterprises

Springer

Philipp Aerni
Center for Corporate Responsibility
and Sustainability (CCRS)
at the University of Zurich
Zürich, Switzerland

ISSN 2191-5504 ISSN 2191-5512 (electronic)
SpringerBriefs in Economics
ISBN 978-3-030-03797-0 ISBN 978-3-030-03798-7 (eBook)
https://doi.org/10.1007/978-3-030-03798-7

Library of Congress Control Number: 2018960733

This Springer imprint is published by the registered company Springer Nature Switzerland AG
The registered company address is: Gewerbestrasse 11, 6330 Cham, Switzerland

Preface

I was in my early teens when my teacher, a member of a radical Swiss left-wing party, encouraged me to ask critical questions about the foundations of our affluence in Switzerland. I agreed with him that our wealth must be related to the widespread poverty elsewhere. In other words, he convinced me that international trade and Foreign Direct Investment (FDI) must be a zero-sum game that merely benefits one party at the expense of another, reflecting the nature of capitalism as a system of exploitation. It also made me believe in our obligation to make up for the damage we cause elsewhere by giving generous financial support for overseas development assistance.

Only when I wrote my Master thesis in Geography in the mid-1990s on the problem of indebtedness of mountain farmers involved in commercial agriculture in Guatemala, I realized that the situation is more complex. Foreign aid is not necessarily the solution but can also be part of the problem, whereas FDI is not necessarily the problem but can also be part of the solution.

Yet, when I presented the findings of my field research to my thesis advisors in Geography and Economics back home, they were not impressed. They especially disliked the fact that, based on my prior inductive field research in Guatemala, I developed my own hypothesis on the circumstances that increase the risk of farm households to become indebted. I empirically tested this hypothesis through a representative farm household survey and data provided by the cooperative that supported the selected farmers. The results were not in line with the theories neither in development economics nor in human geography. I believed that the inconsistency between theory and empirical evidence on the ground was related to the fact that the theories were developed during the Cold War period, when 'development cooperation' still served a concrete purpose of foreign policy, namely to ensure that a client state in the Global South will not suddenly change sides in the global confrontation between the two superpowers, the United States and the Soviet Union, and their respective allies.

The situation in the 1990s was very different because the bipolar turned gradually into a multipolar world order. At the same time, the start of the digital revolution made it easier for low-income countries to catch up, provided that governments created the necessary enabling environment to attract long-term investments from Multinational Enterprises (MNEs) that were also willing to embed themselves into the local economy. The digital revolution also made it easier for watch dog organisations to detect and publicize corporate abuse. The resulting global knowledge-based economy did not necessarily become more stable but it definitely contributed to the economic empowerment of many countries that previously belonged to the very poorest in the world. The economic rise made it possible for many of these emerging economies to not just compete with the established western economies and Japan, but also to pursue their own, economically more pragmatic type of development assistance through South-South Collaboration.

These changes in the real world economy did however hardly affect the curricula in social science disciplines such as economics or human geography. Economics continues to define itself as the science of the optimal allocation of scarce resources even though the resource 'knowledge', the foundation of today's global knowledge economy, is not scarce. Unlike scarce material goods, knowledge is an intangible good that does not decrease but actually increases in value, the more it is used, as Paul Romer, nobel prize winner in economics in 2018, pointed out. Since entrepreneurs are the ones that make use of new knowledge to create innovations that help address scarcity problems, they play a crucial role in enabling sustainable change in the 21st century. Entrepreneurs remain however a black box in economics.

Human Geography, in return, may use a different jargon but continues to be guided by the baseline assumptions of Cold War theories that dealt with the cultural legacy of the former colonial rule to explain 'underdevelopment'. Such theories tend to frame low-income economies per se as 'vulnerable' in the global system of economic exchange. To label its people 'vulnerable' has however the implicit consequence of denying them the role of active economic agents beyond their neighbourhoods. As passive victims of external economic shocks, they must be supported through effective resilience strategies recommended and funded by foreign experts. The result of these well-meant external interventions is often more economic dependence, not less.

So why do these theories remain so popular? Maybe because the assumption that economic globalization is either a zero-sum game or merely creates tiny efficiency gains through trade at the expense of large negative social and environmental externalities, sounds intuitively right. The economic historian and well-known critic of economic globalization Karl Polanyi, who coined the term 'embeddedness' in academic literature, described this negative effects of economic globalization for local people already in 1944 in his book 'The Great Transformation'. His early warnings about the dangers of unbridled market forces for locally embedded economic systems resonate well with many contemporary social scientists concerned with Trumpian authoritarianism, the global systemic risks of capitalism, business and human rights, and social inequality.

Yet, once we recognize that global trade and FDI are not zero-sum games, like war, or, in most cases, the former military-backed neo-colonial rule, Polanyi's anti-globalization narrative appears less adequate to explain the complex challenges we face in the 21st century. All the concerns are real, but they will not go away by simply protesting or by advocating a return old models of social utopia that failed in practice. Instead, there is a need for adjusting and combining social science theories so that they become more responsive to today's global realities; they must build upon a contextual and dynamic understanding of sustainability in order to provide a fruitful theoretical foundation for sustainable collective action designed to not just minimize the external risks but also harness the external benefits of private sector investment for society and the environment.

The present book challenges the simple distinction made by Polanyi and his contemporary disciples between the 'desirable' local versus the 'undesirable' global. At the same time, it proposes an alternative interdisciplinary social science approach that draws on insights from economic sociology and business research. It illustrates under which institutional framework conditions, global business with a commitment to local embeddedness may become less of a risk and more of an opportunity to local economic empowerment and cultural renewal.

Yes, MNEs are free to choose where to invest, and they tend to pick locations where production costs are low. However, many of them are also aware that their investments do not take place in a vacuum. If MNEs only feel accountable to stakeholders elsewhere, they may eventually face a local legitimacy problem. In this context, a commitment to 'principled embeddedness' may help address external concerns about compliance with principles of responsible investment as well as local concerns about inclusive growth. An MNE may generate inclusive growth as a side-effect of its interest to embed itself into the local economy and culture. Its local engagement may also enable the MNE to gain local trust in the form of social capital, which increases its local acceptance and thus helps securing its license to operate.

However, embedding a global company in local business is fraught with cultural misunderstandings as well as economic risk and uncertainty, especially in low income countries. It may also require a significant amount of investment in the upgrading of local skills, capacities and infrastructure, as well as efforts to reconcile local cultural habits with the necessity to comply with the formal rules of global business. In this sense, MNEs alone may not have the means and the competences to succeed; they depend in this endeavour on local partners in government, civil society and the private sector. If these local stakeholders help the MNE succeed in this venture, they also contribute to a considerable amount of external social benefits that result from embedded FDI. These indirect benefits for the local people and their environment go far beyond of what is called 'spillovers' in economics. After all, an embedded MNE does not just bring material goods into the country of destination but also knowledge and know-how as well as access to global networks and capital. These non-tangible resources are an essential condition for inclusive and sustainable change, a central objective of the Sustainable Development Goals (SDGs) of the United Nations (UN). They are a driver of economic integration provided that society supports the selective hybridization of global business and local culture.

A major obstacle to more sustainable global economic integration is the defensive framing of sustainability in postmaterialist societies in general and academia in particular. It is expressed in the popular concern that global economic growth poses a mere threat to local sustainability and human rights. The UN

Guiding Principles on Business and Human Rights (UNGP) illustrates this with its 'do no harm' requirement for MNEs. It ignores that MNEs may also have a potential to directly or indirectly enhance access to essential human rights such as right to food, water, shelter, education, health etc through their local investments. After all, they often represent economic powerhouses in the respective region that provide local jobs and enhance business opportunities for local entrepreneurs. The result is an increase in local income per capita, which automatically enhances access to essential goods and services. Yet, neither the UNGP nor the OECD Guidelines for Multinational Enterprises, that have been translated into national action plans for business and human rights, suggest to balance the potential risks against the potential benefits in human right impact assessments of MNEs in low income countries.

The on-sided focus on risk avoidance results in a global compliance and due diligence bureaucracy mainly concerned with corporate reputation management. The focus on preventing risks tends to discourage investments in local entrepreneurship and innovation in high-risk economies in the Global South that would desperately need more investment. This is however of little concern to anti-globalization activists who envision the restoration of something better that they believe must have existed prior economic globalization. As media savvy 'epistemic brokers' their globally coordinated protest actions provide meaning and orientation by converting complex local stories of resistance into simplified and dramatic mythical accounts of 'big business' versus 'the people'. By doing so, they often misrepresent the demand for inclusive growth that has mobilized local people in the first place. In other words, they deprive local people of their voices while claiming to represent their concerns. With their often divisive and inflammatory language they also discourage effective cooperation between civil society and the private sector, as envisioned by SDG 17 on the creation of a global partnership. The fact that many of these anti-globalization celebrities nevertheless obtain honorary doctorates from social science departments at prestigious universities, is another indication for the existence of a serious crisis of theory in the social sciences.

The book proposes to overcome the popular anti-globalization narrative by telling an alternative story of globalization. A story that is told by the local people in low income countries that have made an experience with global business. This experience may sometimes be positive, and sometimes negative, but it helps to reach a more differentiated view that also restores agency to the local people. Once their voices are heard in postmaterialist societies, it may initiate a reflection process that will also impact the willingness of academic, non-government and government institutions to abandon old but cherished stereotypes and instead embark on pragmatic collective action with the private sector.

The arguments and policy recommendations in this book about the impact of embedded MNEs, the facilitating role of development assistance and the potential of academic research and civil society to better contribute to the SDGs are based on insights from interdisciplinary social science research and illustrated by means of selected business case studies of MNEs operating in low-income countries. They highlight the challenges and opportunities of embedded MNEs and how they can contribute to the achievement of the SDGs, if committed to responsible local investments and supported in their embeddedness strategy by local institutions.

The motivation to write this book is to a great extent rooted in my personal biography as a human being who believes in the moral obligation to ask critical questions, and as an interdisciplinary social science scholar who is alarmed about the trend in the social sciences to return to strictly disciplinary academic careers—even in times when the demand for interdisciplinary knowledge and field research is increasing in society worldwide.

The content and the structure of the book have been significantly influenced and improved thanks to the valuable feedback from Douglas Southgate, Constantine Bartel, Thomas Cottier, Isabelle Schluep, Paul Slovic, Jayashree Watal, and Ron Herring. I would like to thank them all for the proofreading and their very valuable suggestions.

I would also like to thank my mentor Calestous Juma, who sadly passed away in December 2017.

Zürich, Switzerland Philipp Aerni

Contents

1 Introduction 1
1.1 Karl Polanyi's Influence in the Globalization Debate of the 21st Century 3
1.2 The Bipolar Mindset in Academia, Civil Society and Government 4
1.3 Acknowledging the Value of Companies Committed to 'Principled Embeddedness' 6
1.4 When MNEs Become Part of the Solution Rather Than Part of the Problem 8
1.5 Of Myths and Movements 9

2 Societal Foundations of Economic Development 13
2.1 Polanyi as the Common Denominator of Post-structuralism and Neoclassical Economics 14
2.2 No Such Thing as a 'Globalization Paradox' 14
2.3 Polanyi as 'Intellectual Guide' in Economics and Anthropology 18

3 Neoliberalism: A Mythical and Meaningful Term Devoid of Any Deep Thought 21
3.1 Did the Rent-Seeking Economy of Feudalism Serve the Needs of the People? 22
3.2 The Enclosure Movement in the UK as the Beginning of Industrial Agriculture 23
3.3 Blaming Agricultural Trade Has Never Solved Any Food Security Problem 24
3.4 Fernand Braudel's Criticism of Polanyi's Interpretation of History 25
3.5 Why Polanyi's Bipolar Framing Finds Fertile Ground 26

4 The Impact of Popular Stereotypes in Academic Research and Public Policy 29
4.1 Echo Chambers: The Attack on Democracy from Within 30
4.2 Embedded Liberalism: A Flawed Concept 31
4.2.1 The Artificial Separation of the 'Authentic' Local from the 'Generic' Global Product 32
4.2.2 Governments as the Blameless Defenders Against Careless Big Business 33
4.2.3 Why More Regulation Does not Lead to More Public Trust: The Case of GMOs 34
4.2.4 Self-Regulation in Industry as a Base for Subsequent Government Regulation 35
4.2.5 Why the Ruggie Framework May Not Be Harmless 35
4.3 Indigenous Communities as Projection Screens for Preserved Cultural Embeddedness 36
4.3.1 How the Indian Chipko Movement Became a Symbol of NIMBY Environmentalism 38
4.3.2 Myths Embodied in Scientific Models that Guide Academic Research 41
4.3.3 'Epistemic Brokers' in Postmaterial Societies: The Case of Vandana Shiva 43
4.3.4 Cultural Appropriation and Denial of Local Agency 46
4.3.5 The Temptation in Academia to Uncritically Embrace Environmental Narratives 48

5 The New Understanding of the Term 'Embeddedness' in Economic Sociology 51
5.1 The Moral Dimension of Entrepreneurship 52
5.2 Embeddedness as a Way to Address Three Major Coordination Problems 52
5.3 Value as a Coordination Problem 53
5.4 Competition as a Coordination Problem 54
5.5 Cooperation as a Coordination Problem 56
5.6 Embeddedness in the Context of Economic Complexity 58

6 Economic Globalization as a "Disembedding" Force? 61
6.1 Why Disembedding Traditional Structures May Help Outsiders 62
6.2 Disembedding Post-Colonial Structures 63
6.3 How FDI Contributed to Catch-up Growth and Economic Empowerment 64
6.4 'Knowledge', an Underused Resource in Efforts to Cope with Environmental Challenges 66
6.5 The Failure of Foreign Aid to Empower Local Entrepreneurs Through Economic Integration 67

6.5.1 Self-Serving Nature of Swiss Sustainable Trade Promotion and Development Research 68
6.5.2 How the Anti-business Rhetoric of Epistemic Brokers Supports Incumbents 71
6.5.3 Local Growth-Oriented Entrepreneurs as Drivers of Economic Integration 71

7 Embedded MNEs and Their Contribution to Sustainable Change 73
7.1 Coping with Business Coordination Problems Through a Strategy of 'Principled Embeddedness' 74
7.2 The UNGP and Its Potential Conflict with Principled Embeddedness 75
7.3 The Role of Subsidiaries of MNEs in Developing Countries 76
7.4 Selected Cases of 'Principled Embeddedness' of Subsidiaries of MNEs 77
7.4.1 Nestlé Philippines: An Locally Embedded Company 78
7.4.2 Syngenta's Contribution to Capacity Development for Agricultural Innovation 79
7.4.3 Bata Shoes: Creating Welfare by Taking Rather Than Avoiding Risk 81
7.4.4 The Responsible Entrepreneur and the Selfless Communist 83
7.4.5 Chiquita: A Pioneer in Sustainable Banana Production with a Legacy Problem 84
7.4.6 The Problem with Business to Consumer Labels in Agriculture 86
7.5 Embeddedness and Its Link to Sustainability and Corporate Responsibility 87

8 Development Cooperation as a Catalyst for Sustainable Long-Term FDI 89
8.1 How Development Assistance (DA) Could Encourage Foreign Direct Investment (FDI) 91
8.1.1 Sourcing Ingredients for Beer Production in Uganda 92
8.1.2 Empowering Pastoralists Through Business Development in Kenya 93
8.1.3 Enabling Access to Finance: The Case of Vodafone 94
8.2 Lessons Learned from Public-Private Partnerships (PPPs) 96

9 Concluding Remarks 99

References 111

Acronyms

ACRE	Agriculture and Climate Risk Enterprise (ACRE)
AoA	Agreement on Agriculture
ATDF	Africa Technology Development Forum
B2B	Business to Business
B2C	Business to Consumer
BSCI	Business Social Compliance Initiative
CCRS	Center for Corporate Responsibility and Sustainability
CIME	OECD Committee on Investment and Multinational Enterprises
CORBANA	Corporación Bananera Nacional Corbana
COSA	Committee on Sustainable Agriculture
CSR	Corporate Social Responsibility
CTI	Swiss Federal Commission on Technology and Innovation
DA	Development Assistance
DFID	UK Department for International Development
ESG	Environment, Social, Governance
ETH	Swiss Federal Institute of Technology
FAO	Food and Agriculture Organization of the United Nations
FDI	Foreign Direct Investment
FOAG	Swiss Federal Office of Agriculture
GBF	Grain Bulking Facility
GDP	Gross Domestic Product
GI	Geographical Indication
GiZ	German Group for International Cooperation
GM	Genetically Modified
GMOs	Genetically Modified Organisms
GVC	Global Value Chains
ICIJ	International Consortium of Investigative Journalists
ICT	Information and Communication Technologies
IFPRI	International Food Policy Research Institute
IMF	International Monetary Fund

IP	Intellectual Property
ISO	International Organisation for Standardization
MDGs	Millennium Development Goals
MNC	Multinational Corporation
MNE	Multinational Enterprise
MRV	Measurement, Reporting, Verification
NBL	Nile Breweries Limited
NCPs	National Contact Points
NGO	Non-Government Organisation
NIMBY	Not In My Backyard
NRP	National Research Program
OECD	Organisation for Economic Co-operation and Development
ORT	Oral Rehydration Therapy
PPP	Pubic Private Partnership
R4D	Research for Development
RA	Rainforest Alliance
SAN	Sustainable Agriculture Network
SDC	Swiss Development Cooperation
SDGs	United Nations Sustainable Development Goals
SECO	Swiss State Secretariat for Economic Affairs
SMEs	Small-and medium-sized Enterprises
SNSF	Swiss National Science Foundation
TAP	Tropical Agriculture Platform
TFA	Trade Facilitation Agreement of the WTO
TFP	Total Factor Productivity
TRIPS	Trade-Related Aspects on Intellectual Property Rights
UNCTAD	United Nations Trade and Development Conference
UNDP	United Nations Development Programme
UNGP	United Nations Guiding Principles on Business and Human Rights
UP	Uttar Pradesh
USAID	United States Agency for International Development
UVAMA	Uganda Value Added Maize Alliance
VAC	Village Aggregation Centers
WEIRD	Western, educated, industrialized, rich and democratic
WHO	World Health Organisation
WTO	World Trade Organisation
WWF	World Wild Life Fund

Chapter 1
Introduction

The famous economic historian Karl Polanyi called the expansion of formal global markets in his seminal book 'The Great Transformation' (1944) a 'satanic mill' that would lead to the disembeddedness local communities and their informal economies. Polanyi's framing of economic history as a struggle of 'profit versus people' has gained renewed attention in the 21st century.

Robert Kuttner, an American journalist and social policy expert, argues in a recent essay in the New York Review of Books (Kuttner 2017) that austerity policies in Europe and the renewed push for deregulation in the United States would reaffirm what Polanyi criticized as "the utopian endeavour of economic liberalism to set up a self-regulating market system". This endeavour would crowd out local culture and citizenship; lead to extreme inequality, and eventually trigger a political counter-movement to restore human rights to ordinary people.

This book does not defend European austerity policies or the Trump administration's fondness for deregulation of the domestic economy—while simultaneously rejecting multilateralism in the governance of the world trading system. It does also not deny the countless corporate scandals before and after the global financial crisis of 2007–2008 that ruined the lives of many ordinary people. Yet, it challenges the popular narrative of global business as a sort of zero-sum game that merely thrives at the expense of society and the environment. In a world characterized by a high degree of economic interdependence, social and geographical mobility and transboundary environmental and social challenges, global business cannot be regarded anymore as something external and alien that is unrelated to our personal lives and social networks. We are all directly or indirectly dependent on and also benefit from its products, services and innovations as local producers as well as local consumers. The global sustainability challenge of the 21st century is therefore not to get rid of global business but to better harness its potential to contribute to local sustainable development and inclusive growth.

Multinational Enterprises (MNEs) are the main players in global business and probably the most scrutinized ones. The belief that their global operations are completely detached from local cultural and social activities contradicts the fact that

P. Aerni, *Global Business in Local Culture*, SpringerBriefs in Economics,
https://doi.org/10.1007/978-3-030-03798-7_1

all economic relations, whether global or local, are based on prior social relations. In other words, the local cultural dimension should not be pitched against the global economic dimension but must instead be seen as its foundation. There is, of course, corporate culture and there is local culture. However, these are not terms that describe a steady state but represent dynamic processes that thrive on exchange. Embedded foreign investments may contribute to a fruitful exchange by responding to local concerns and by creating new local economic opportunities through economic integration. If MNEs with a commitment to principled embeddedness[1] succeed in becoming an accepted and respected player in the local economy and culture, these companies gain the necessary social capital to secure their long-term license to operate. In other words, corporate social responsibility (CSR) is built into the long-term interest of such firms. CSR thus ceases to be a separate section with a separate agenda within the MNE. Instead, it becomes an integral part of an overall business strategy designed to ensure the long-term survival of the company within society.

The potential contribution of such MNEs to sustainable change in developing economies is seldom appreciated because it runs counter to the stereotype that MNEs merely privatize profits while socializing the costs in the regions where they invest.[2]

The general view that global companies do business at the expense of local cultural and economic activities remains firmly entrenched in the sustainability debate of affluent societies. It often leads to an exclusive concern for the defense of the local, the search of a like-minded online and offline communities and the withdrawal form an engaged political debate. The mentality tends to be captured by the term 'Not in my Backyard' (NIMBY), which has the unintended consequence of encouraging cultural segregation rather than economic integration. Yet, the bipolar view of the 'bad' global and the 'good' local that underpins this static and defensive view of sustainability runs counter to the Sustainable Development Goals

---

[1]'principled embeddedness' stands for the corporate commitment of an MNE to follow its self-imposed corporate responsibility principles worldwide, while, simultaneously, providing its subsidiaries with sufficient autonomy to embed themselves into the local economy (see Aerni 2017b). In this sense, the term 'principled' ensures that embeddedness is not indirectly endorsing the undesirable type of embeddedness associated with local corruption and collusion (an anti-corruption policy should be part of the CSR principles of a company).

[2]There are certainly global companies that do indeed only care about profits no matter at what social cost. Moreover, the so-called 'paradise papers', published on November 5, 2017 by the International Consortium of Investigative Journalists (ICIJ) have once again unveiled tax evasion practices by MNEs, especially in the mining sector, that may not be illegal but nevertheless account for a significant loss of government revenues in countries where the resources are extracted. Abusive tax practices by MNEs will be addressed briefly in Sect. 7.4.4. In view of the ease to hack data in the age of the digital society and the enormous long-term costs resulting from reputation loss associated with the detection and denouncing of extensive transfer pricing, MNEs may think twice if it is worth to endanger their license to operate by taking the risk of making extensive use of tax havens to optimize tax payments. Some MNEs may continue to engage in tax evasion even if they cannot claim that national corporate taxation schemes would be abusive. But it would be unfair to argue that they represent MNEs in general.

(SDGs) who aim to promote inclusive growth (SDG 8) and, for that purpose, call for a global partnership (SDG 17) to develop hybrid and tailor-made local solutions to effectively address the significant global sustainability challenges of the 21st century.

1.1 Karl Polanyi's Influence in the Globalization Debate of the 21st Century

Economic globalization is primarily associated with the growth of multinational enterprises (MNEs). They have their headquarters primarily in prosperous economies in North America, Europe, and Asia and focus increasingly on investing in developing countries where land and labor are still relatively cheap.

Economists and political scientists who represent the school of 'Embedded Liberalism' (Ruggie 1982; Hays 2009; Rodrik 2011) and scholars in the field of 'Corporate Social Responsibility' (CSR) (Scherer et al. 2006; Wettstein 2010) regard this trend as potentially disruptive for traditional communities and their locally embedded economic systems. Weak law enforcement capabilities in developing countries would be unable to ensure the protection of human rights of such communities and therefore public and private initiatives are required to compensate affected communities through a generous welfare state or global CSR initiatives, respectively.

The view that the primary task of governments is to tame unfettered market forces is not necessarily wrong but incomplete and often accompanied by an ideological agenda and vested interests in preserving the status quo. On the left wing of the political spectrum, anti-globalization activists demand additional regulation designed to minimize the social and environmental risks of global corporate investment. On the other side of the political spectrum, right-wing nationalists frame global economic integration and migration as threats to cultural identity and national sovereignty. The recent political successes of the far right in North America and Europe have been made possible because of the support of the 'distributional losers' of globalization and the sedentary middle class that is concerned about cultural and economic decline. These losers feel increasingly decoupled from global economic change and no more represented by the left wing politicians that have shifted their concerns from the domestic worker to 'vulnerable minorities' (Hopkin 2017; Pepinsky 2017; Reckwitz 2018). They ask for simple explanations to complex problems, and political entrepreneurs, who play the role of epistemic brokers, provide such explanations by using popular narratives of 'good' and 'evil' forces, and by identifying plausible scapegoats (Aerni and Bernauer 2006).

Yet, the claim that an unfettered global economy is disrupting the local economies in an unprecedented way does not correspond to today's reality of mixed economies with their subsidies and policy interventions to protect the domestic economy from world trade (Rogers 2017). Such protectionist policies, especially when combined with non-tariff trade barriers, are often justified by the almost unquestioned chauvinistic assumption that everything produced domestically is automatically more sustainable and of better quality than substitutes produced abroad. Such protectionist policies often favour potent incumbents in domestic business. They primarily aim at preserving the status quo by arguing in favour of protecting the 'embedded' national economy, understood as a highly regulated economic system that protects the local business against disruptive economic change driven by entrepreneurship and innovation. For outsiders, within and without the domestic economy, who do not benefit from the social network and the political connections of incumbents in the resulting corporatist system, such an 'embedded' economy is primarily characterized by nepotism (Schluep and Aerni 2016). It stifles their economic opportunities. Therefore, entrepreneurial outsiders see economic globalization not just as a threat, but also an opportunity to weaken the dominant position of incumbents in domestic economies and make space for more economic freedom. The chances of such outsiders to find ways around established networks and create new and scalable markets has increased with the digital revolution and the rise of the global knowledge economy (Naam 2013). Yet, public resentment against such agents of change persists, especially when they become successful and grow big.

1.2 The Bipolar Mindset in Academia, Civil Society and Government

The new opportunities offered by the global knowledge economy of the 21st century require a critical re-evaluation of Polanyi's dualist worldview that guided his interpretation of economic history in the first half of the 20th century.

The re-evaluation of Polanyi takes place in the second, third and fourth chapter, as well as Sect. 7.3 of the present book. It builds upon existing research in economic history (Braudel 1982; Stehr 2008; Romer 2010; Bang 2016), economic sociology (Granovetter 1985; Zafirovski 2002; Beckert 2007), and industrial policy (Uzzi 1996; Meyer et al. 2011). This empirical research challenges the implicit baseline assumption of the school of 'embedded liberalism', which is based on Polanyi's argument that the global expansion of the formal market system poses an exclusive threat to locally embedded economic systems and human rights. The claim of Polanyi's contemporary disciples that the World Trade Organisation (WTO) merely represents the interests of the global actors at the expense of local interests is shown to be misguided in Chap. 2. After all, the WTO is a product of compromise. It reflects the wish of its member states to participate in a rule-based economy with ample policy space that also takes into account non-trade concerns other than rent seeking protectionism. This policy space, incorporated in the different WTO Agreements, is especially significant for member states that belong to the category of Least Developed Countries. Some global activists would object by pointing out that many interest groups were underrepresented in the negotiations of the WTO Agreements. Indigenous people, for example, would feel threatened in their cultural identity by the expansion of global business encouraged by the WTO. Section 4.3 contradicts this view by pointing out that the local interests of indigenous people are often misrepresented by the global civil society organizations that claim to represent them on the global stage. After all, Article 21 of United Nations Declaration on the Rights of Indigenous Peoples from 2007 clearly demands respect for the economic rights of indigenous people (right to ownership, right to self-determination and economic development, equality before justice, freedom from discrimination). It indicates that indigenous people are as much interested in fair economic and cultural exchange as they are in cultural preservation. They are aware that their indigenous culture only remains attractive to the next generation, if there is cultural renewal supported by selective economic integration.

Despite the lack of empirical evidence, the belief that the expansion of the formal global economy goes at the expense of local people who defend their local culture and natural environment has become very popular in contemporary affluent societies since they are mainly interested in preserving the status quo, from which they benefit. Since their material needs are well taken care of, they are largely concerned with postmaterial needs, especially the search for identity and meaning. In this context, the mythical accounts of 'local people' versus 'global profit' spread via social media by well-known anti-globalization activists, such as Vandana Shiva, resonate well with them. Their media-covered stories are often framed as a 'David versus Goliath' drama and, as such, provide meaning and orientation. The narrative may sound simple and plausible, but it is highly misleading. After all, no real profit-oriented economic activity is detached from individuals (people) and their social networks (communities), as economic sociologists have pointed out in their embeddedness research.

The embeddedness in social networks is especially crucial when a company invests abroad. It must gain social capital in the region of investment by contributing to the resolution of three major coordination problems: the problem of value, the problem of competition and the problem of cooperation. Chapter 5 points out that foreign investors may only be able to secure their long-term license to operate in the host country if these coordination problems are adequately addressed in collaboration with local stakeholders.

Chapter 6 looks at contemporary economic and development policies that are still guided by the social science theories developed during the Cold War. Structuralist and neomarxist theories, that were very popular in Human Geography and Postcolonial Studies, implied for example that international trade must be, just like conquering and subjecting countries, a zero-sum game that benefits the rich at the expense of the poor. In turn, neoclassical economics, also a theory developed during the Cold War period, only focuses on the relatively modest efficiency gains from global trade while ignoring the welfare effects generated through the introduction of new goods and services (Romer 1994). Moreover, welfare economics, a branch of neoclassical economics, has an exclusive focus on internalizing the negative externalities caused by private sector activities. The positive external effects on society resulting from private sector investment in innovation are largely ignored.

The expansion of the global economy after the Cold War has however significantly benefited previous low-income countries, such as China. By carrying out the institutional reforms necessary to embark on catch-up growth, China was less focused on capturing efficiency gains from trade but on taking full advantage of the economic opportunities resulting from the rise of the global knowledge economy.

The global knowledge economy is strongly linked to the ongoing digital revolution that made the non-rival resource 'knowledge' more widely available. Yet, access to codified knowledge on the internet does not yet ensure development. The more important part is investment in human capital to create the necessary tacit knowledge (know-how) to make commercial use of codified knowledge. In this context Foreign Direct Investment (FDI) is very valuable because it combines knowledge with know-how transfer into the local economy. These essential ingredients of endogenous development increase the likelihood of imported physical goods to be eventually substituted by locally produced goods. Being a non-tangible resource, knowledge in the form of instructions, recipes, and protocols makes it possible to create a local good that is otherwise too costly to import—provided that the country has invested in the business infrastructure and the human skills and know-how of its people to take advantage of the new opportunities.

In this context, the effectiveness of Official Development Assistance (DA), which was also invented during the Cold War to win over non-aligned developing countries, is increasingly questioned. It is stuck in the classic view that development aid must primarily protect rather than economically empower the poor (Easterly 2007; Deaton 2015). As such, DA tends to preserve unsustainable local structures in low income countries rather than enable the highly needed structural change to create new economic opportunities for the large and increasingly educated younger generations in the developing world.

1.3 Acknowledging the Value of Companies Committed to 'Principled Embeddedness'

Chapters 7 and 8 argue that effective DA needs to build upon the principle of cooperation, especially with the private sector, if its goal is to enable sustainable change by reducing poverty through more economic opportunities and, simultaneously, improving the environment through sustainable intensification. Producing more with less by making effective use of new platform technologies such as information technology, nanotechnology, and biotechnology, is vital in view of global population growth and increasing affluence in the 21st century. In this context, the focus in public policy and CSR needs to shift from merely regulating and avoiding the risks of FDI to harnessing its benefits for the poor and the environment. It must be based on the insight that investments of MNEs do not just cause external costs for the local environment and society but may also generate external social and environmental benefits, especially if the MNE is committed to 'principled embeddedness'.

In this context, Chap. 7 points out that the UN Guiding Principles on Business and Human Rights (UNGP)[3] as well as various other international CSR guidelines developed by the Organisation for Economic Cooperation and Development (OECD) and the International Organisation for Standardization (ISO), should recognize that corporate responsibility cannot be limited to the requirement of doing 'no harm'. After all, companies do not create value by merely avoiding risks but by actually taking the risk to invest in a new market. MNEs that benefit the region in which they operate should also be rewarded for doing 'good,' not because they want to be good corporate citizens but because doing so is in their long-term interest.

A possible reason for the omission of the importance of local embeddedness in current CSR strategies may be the influence of global retailers on the design of sustainability standards in the food and agricultural sectors. Rather than informing consumers about the efforts of the supplying global agribusiness companies to make agriculture in developing countries more sustainable, they prefer to portray themselves as the most sustainable companies in the global food value chain. Section 7.4.6 illustrates how they do so by informing consumers about their collaboration with reputation-enhancing environmental organizations, such as WWF. In their marketing campaigns, they primarily aim at making consumers feel good about themselves and confident about their retailer as a selfless defender of nature and small-scale farming (Miller 2012; Aerni 2013a). The essential pillars of this wellness sustainability are 'organic' or 'fair trade' premium products portrayed as natural, healthy and fair and therefore a more ethical alternative to industrial agriculture.

These claims are increasingly questioned based on insights gained from empirical research (Makita and Tsuruta 2017; Huybrechts et al. 2017; Laufer 2014; Lott 2015; Ramone 2013; Gilbert 2012; Henderson 2008). Even from an ethical point of view, it is unclear whether these wellness premium products are the best choice for consumers. There is increasing evidence from field research, that 'fair trade' and 'organic' production in developing countries may help increase the income of the immediate beneficiaries (e.g. members of the respective farm cooperative) but discourage local entrepreneurship and innovation, the key ingredients for homegrown development. Moreover, 'fair trade' and 'organic' cooperatives in low-income countries are controlled by retailers in high-income countries. As such, they tend to be capital-intensive tropical food production sites subsidized by foreign consumers and states, but in most cases, utterly disembedded from local

[3]See http://www.ohchr.org/EN/Issues/Business/Pages/BusinessIndex.aspx (Accessed Sept 25 2018).

economic activities. As such, these niche markets for affluent consumers contribute very little to structural change.

Finally, it is unclear whether ethical concerns indeed motivate consumer decisions to buy organic or fair trade, or whether it is rather about treating oneself to a premium product (Miller 2012). No one would probably be puzzled if the marketing slogan for these 'sustainable' products would be 'because I am worth it'.

1.4 When MNEs Become Part of the Solution Rather Than Part of the Problem

Whether FDI is indeed capable of generating profits by empowering rather than exploiting people depends on the awareness of the MNE that its business does not operate in a vacuum as well as on the respective domestic institutional setting. Governments that want to force foreign investors to comply with local content requirements may not be effective in achieving the desired outcome if they fail to "do their homework." This homework comprises domestic reforms and investments, not just in the domestic human capital stock and an institutional environment that enables economic and technological change, but also in the physical and digital infrastructure (UNCTAD 2017a).

Creating these favourable framework conditions helps reduce the uncertainty for subsidiaries of MNEs to invest in the domestic economy.

The commitment to 'principled embeddedness' in practice is illustrated in this book in Sect. 7.4 using selected MNE case studies. In addition, Chap. 8 uses concrete case studies in Africa to highlight the importance of state and non-state actors as intermediaries and catalysts who render local institutions and businesses more responsive to MNEs prepared to source more of their services and products locally and thus become more embedded.

These case studies are not meant to praise the selected MNEs for their local business practices, but instead to point out that the long-term profit-motive may not necessarily conflict with social and environmental objectives. The potential for opportunism in large companies may be widespread despite increasing expenses on compliance and due diligence processes (Chen and Soltes 2018). However, the examples clearly show that globally active corporations may contribute to economic empowerment and sustainable change in the regions of investment through a strategy of principled embeddedness. This is particularly true for their presence in many developing countries where they often offer an alternative to discriminating traditional economic systems, in which social status and not individual merits determine one's chances to obtain a decent job in the formal economy (Martin 2012).

The ambitious young and educated majority in developing countries who are stuck in persistent feudalist structures tend to become outsiders in their own society. Lacking the necessary social connections to enter the formal economy as entrepreneurs, they often decide to migrate elsewhere in search for economic opportunities. In this context, the growing number of economic refugees is a symptom of failed domestic policies as well as misguided development assistance (DA) that avoids productive collaboration with the private sector to create economic opportunities for entrepreneurs in the formal domestic economy. The grievance of the neglected young entrepreneurs has been identified as one of the main triggers of the Arab spring (Martin 2012).

It is not surprising that Karl Polanyi never addressed this downside of traditional economies in which land-owning insiders officially portray themselves as custodians of the natural environment, traditional society, and local culture to strengthen the legitimacy of their privileged economic and political position in society. It would have conflicted with his binary thinking of the 'bad' global versus the 'good' local. This binary thinking is however persistent because it offers a reduction of complexity in an ever more complex society (Luhmann 1993). As a result, the bipolar mindset continues to influence the funding priorities in development cooperation, academic research and CSR strategies, despite the growing empirical evidence that it aggravates rather than resolves the social and environmental sustainability challenges of the 21st century. After all, focusing only on the 'vulnerable', the presumably passive victims of change, and portraying the agents of change, understood as foreign direct investors and local entrepreneurs who strive to become their suppliers, as mere perpetrators, supports the interests of the incumbent elite rather than the poor who aim to improve their economic situation. Incumbents, after all, benefit from the status quo.

The concept of vulnerability treats the target population as passive victims who need to be saved by 'therapeutic entrepreneurs' supported by DA (Ecclestone 2017). These therapeutic entrepreneurs are assumed to be better educated and therefore to know better what the vulnerable need. In most cases, these external stakeholders are not aware that their interpretation of the local circumstances is in most cases not informed by the local people and their concerns, but somewhat static stereotypes and mythical stories that prevail in donor countries about the situation of the poor in recipient countries. An issue that is extensively discussed in Chap. 4 as well as the concluding remarks of Chap. 9.

1.5 Of Myths and Movements

The view that entrepreneurs who try to take advantage of economic opportunities are mere perpetrators who do not need any assistance proves to be one of the most widespread myths in affluent societies. Why? Because, worldwide, the self-employed, in most cases survival entrepreneurs, live in a much more precarious state than those with formal employment. Moreover, this is not just true for daily laborers without any formal education or training but also university graduates in low-income countries who do not have the opportunity to enter into a family business and failed to obtain a well-paid job with a foreign NGO, an MNE or the government after graduation (Aerni 2015b).

Since the human rights movement emerged from the labour rights movement, the grievances of these entrepreneurs is not on its radar screen (Aerni 2015b). Yet, survival entrepreneurs, especially if endowed with a good education and business training, may be of great interest to foreign companies, which care about motivated and qualified local partners and employees. More than anyone else, these companies give entrepreneurs in precarious situations a chance by investing in their skills or the upgrade of their business. As such, MNEs may significantly contribute to social mobility in traditional societies and the economic empowerment of outsiders.

MNEs in affluent societies are however hardly ever associated with economic empowerment in developing economies. Instead, they are perceived as selfish actors that care about shareholder value, competitive off-shore employment, tax evasion schemes and monopoly power. This may be true for MNEs involved in corporate crimes and malpractices. Hollywood movies and the media widely cover these cases. However, empirical research indicates that the vast majority of MNEs do not correspond to the negative stereotype of 'Big Business'. The claim, for example, that small businesses would account for a higher share of decent employment, be more innovative and contribute more to the tax base of society has mostly be rebutted (WTO 2016; Atkinson and Lind 2018).

Moreover, even though 'big business' may spend more on political lobbying, they seem to be less effective in achieving their goals than small businesses, who generally pay less taxes and obtain much more government assistance (Atkinson and Lind 2018). The reason for this outcome is that public opinion loves 'small' and hates 'big' business; and politicians who care about re-election are careful about not being associated with the 'hated' ones. Alas, by asking for more regulation of innovation-driven industries, the same politicians may inadvertently strengthen the market position of large firms that, unlike small companies, have the means to comply with additional costly regulation (Aerni 2015b). The rather artificial divide between small 'good' firms and bad 'large' firms leads to short-termism in politics that is unable to address the long-term challenges of sustainable development in a collaborative way. After all, small companies must become part of a business ecosystem that also involves large companies, if they want to succeed. Moreover, big companies are probably the largest investors in innovative small companies (Atkinson and Lind 2018).

Given the urgency to move away from the unproductive binary mindset in academia, civil society, and politics and to learn from the past when addressing the global sustainability challenges of the 21st century, Chap. 9 concludes by calling for a paradigm shift in the theory and practice of international sustainable development. This paradigm shift is reflected in SDG 8 on 'Decent work and Economic Growth' of the United Nations Sustainable Development Goals. The United Nations Development Program (UNDP) considers this goal designed to promote inclusive growth to be key for achieving most of the remaining Sustainable Development Goals because improved incomes lift people out of poverty and automatically improve access to essential human rights (e.g., the right to food, right to water, right to decent shelter, gender equality). As such, SDG 8 represents the priority of poor people in developing countries who have not obtained formal employment in the private sector and are therefore forced to make a living as survival entrepreneurs. To understand why they value FDI if embedded in the local economy, one has to listen directly to the poor in these countries rather than to the anti-globalization activists who claim to talk on behalf of their interests. The concluding remarks illustrate this by using a concrete example of good investigative journalism. It listens and gives voice to the locals and, as such, takes local knowledge and experience seriously. The example shows that, for the poor, it is obvious, that poverty has no cause. It is merely the absence of prosperity, as the interdisciplinary social scientist Jane Jacobs noticed. Alas, concerned people in affluent societies still regard their prosperity as being rooted in poverty elsewhere. It is important they realize that economic exchange, unlike war, is not a zero-sum game.

Chapter 2
Societal Foundations of Economic Development

Great interdisciplinary sociologists such as Emile Durkheim, Max Weber and Norbert Elias and economists such as Joseph Schumpeter, Arthur Lewis, and Albert O. Hirschman never heeded boundaries in the social sciences. For a long time, the primary objective in each of the two disciplines was to understand how societies and their economic activities evolve over time and in space. The social sciences were therefore recognized as a science that deals with irreversible processes (Diamond and Robinson 2010; Diamond 2014). It was equally clear to these interdisciplinary scholars that all economic activities are embedded in prior social networks that form in response to specific challenges and opportunities. Alas, the neoclassical paradigm in economics as well as structuralism in sociology, anthropology, and human geography were less interested in the historical dimensions of human interaction and focused instead on uncovering allegedly universally applicable laws and structures of social and economic exchange (Braudel 1982; Rangan 2000). This also led to a tacitly approved division of labor with economists studying economic phenomena and sociologists studying social phenomena (Beckert 2007; Zafirovski 2002).

Economic sociology, however, has enjoyed a revival since the end of the Cold War with the rediscovery and reformulation of 'embeddedness' as a term that describes how societies evolve. Before this revival, the term has largely been associated with Karl Polanyi's understanding of embedded economies in premarket societies. The assumption that formal economic transactions today have ceased to be guided by social and kinship obligations and therefore lack embeddedness was challenged by the insights gained from empirical social science and business research (Granovetter 1985; Biggart and Beamish 2003; De Bakker et al. 2013). The research uncovered the social networks, online and offline, that provide the very basis for formal business relations on the local as well as the global level. Public resistance against and fear of global economic integration has nevertheless grown, resulting in a global normative discourse on 'embeddedness' that is profoundly defensive in nature and shaped by rhetoric that very much follows Polanyi's previous line of argumentation.

P. Aerni, *Global Business in Local Culture*, SpringerBriefs in Economics,
https://doi.org/10.1007/978-3-030-03798-7_2

2.1 Polanyi as the Common Denominator of Post-structuralism and Neoclassical Economics

When Polanyi published ‘The Great Transformation’ in 1944, one of his main purposes was to demonstrate that 19th-century economic liberalism eventually resulted in fascism in the 20th century. In the 21st century, his grand narrative is being re-applied to show how deregulation of financial markets, trade liberalization, and economic austerity after the end of the Cold War have led to the rise of right-wing populism and plutocracy (Kuttner 2017). The baseline assumption is that institutions of embedded traditional economies are being destroyed by a self-regulating market economy, in which the market no more serves the needs of the people but rather reversed, that people would have to serve the needs of the market. A similar line of argumentation was subsequently used to explain why modern technology has ceased to serve the needs of people (Ellul 1990).

Polanyi’s view is also associated with the term ‘substantivisim’ in economic anthropology and has become one of the foundations of post-structuralist social science, which is essentially based on the ‘economy—society opposition’ understood as the difference between the lived experience of market society and its discursive field (Holmes 2013). While the substantivist ‘embedded’ economy is assumed to be focused primarily on social needs and self-sufficiency, the formalist ‘disembedded’ economy would thrive merely on the rational pursuit of profits while allegedly dispensing with social norms. It presupposes an autonomous formal economy without social content and without any history of its own, an assumption commonly found also in the formal theory of neoclassical economics (Gemici 2008).

2.2 No Such Thing as a ‘Globalization Paradox’

In his book the ‘Globalization Paradox’, published in 2011, the neoclassical economist Dani Rodrik draws on Polanyi’s pessimistic view regarding the disruptive social consequences of economic globalization and regards the resistence of its potential losers in domestic economies as an effective counterforce. Those who fear to become victims of globalization would make use of their democratic rights to defend their interest in national sovereignty and to impose restrictions on the global expansion of markets (Rodrik 2011). Rodrik endorses economic populism as a sort of reaffirmation of the priority of the local against the global power. He believes this to be an adequate response to the growing power of MNEs in shaping the agenda of international trade negotiations, patent rules, international investor tribunals, and independent regulatory agencies, All of which would result in global regimes that disproportionately benefit capital at the expense of labor (Rodrik 2018). In other words, the globalist agenda would pose a risk to labor in locally embedded economies. However, as mentioned earlier, these claims stand in contrast with empirical research indicating that MNEs may contribute substantially to local employment (in part due to the simultaneous expansion of small suppliers to large MNEs), pay better salaries and provide more job security (WTO 2016; Atkinson and Lind 2018).

Rodrik based his argument on the fictitious assumption that governments have always been concerned with the provision of public goods while the private sector would only be concerned with profits through the production of private goods. The fact is, however, that the provision of most public goods started with pre-competitive private sector collaboration. Companies recognized the need for self-regulation to enforce quality standards even before there was public regulation (Ping 2011). The standards were enforced by excluding cheaters from the concerned market. These measures were crucial for maintaining public trust, a good that provides legitimacy in society and ensures low transaction costs (Desai 2003).

Rodrik's 'Globalisation Paradox' (Rodrik 2011) argues that countries in the 21st century cannot have national sovereignty, hyper-globalization, and democracy simultaneously. Instead, they would only be able to choose two out of the three. This hypothesis may sound intuitively right, but it is wrong from a historical perspective. It says more about the comparative static mindset of neoclassical economists than the historical struggle between the rulers and the ruled which manifests itself in the private sector as much as in the public sector. The priorities of the ruled have always been related to economic rights. Unlike the ruling elite (the master) that prefers to stick to the status quo, the ruled want change because they do not like to remain the servants of the ruled (Kojève 1975). Therefore, if globalization creates more economic opportunities and helps dismantle a highly protected domestic rent-seeking economy that only benefits a tiny elite, the servants will endorse economic globalization because it offers a pathway to economic empowerment.

Consequently, for the poor servant class in rent-seeking economies which is economically highly dependent on the ruling elite, economic rights matter at the beginning more than political rights (Aerni 2015b). After all, political rights become consequential only after the ruled have achieved economic empowerment (Sandefur 2010). As economic empowerment has reached an advanced stage, the ruled become the tax-paying entrepreneurial middle class that ceases to be economically dependent on the ruling rent-seeking elite. As such, they start to insist on the right to participate in the political decision-making process. An economically empowered middle class is, therefore, an essential condition for a stable democracy (Aerni 2015b).

The initial focus of the ruled on economic rights may be a reason why people in many parts of the developing world accept authoritarian regimes as long as they provide economic opportunities. China may not be a democracy today, but which European power during the industrialization in the 19th century could have claimed to be a genuine democracy? Back then, Europe faced economic, social and environmental challenges related to accelerated structural change similar to what developing countries face today. In both cases, China in the 21st century and Europe in the 19th century, strong political leadership that was supportive of an experimental approach to find practical solutions to address these challenges was crucial in enabling sustainable change that ultimately ensured that ordinary people benefited from economic and technological change in the long run. This pro-active and progressive way to address public good problems had no relation to Rodrik's contrived dilemma between democracy, sovereignty, and global trade, but was instead based on pragmatic public-private partnerships designed to 'get things done' (Aerni 2015b).

When Rodrik claims that the transition from the General Agreement on Tariffs and Trade (GATT) to the creation of the World Trade Organisation (WTO) in January 1995 represented a disruptive departure from prior flexible trade arrangements that were guided by national interest, to a set of enforceable trade rules that lack democratic legitimacy and undermine national sovereignty (Rodrik 2011), he ignores a wide range of facts. First of all, the WTO cannot be conceived as an external institution that imposes its agenda on nation states. It is rather a club to which 164 countries currently belong that have agreed to subject international trade to a rule-based system rooted in two fundamental principles (Most Favoured Nation, National Treatment) designed to avoid discrimination. The principles of non-discrimination carefully balance market access with the need to address non-trade concerns. In this context, the WTO agreement contains many forms special and differential treatment and safeguard mechanisms, that protect the interests of low income countries. It also provides high income countries with many policy instruments (sensitive products, green box, precautionary principle etc) to raise non-tariff trade barriers and protect entire economic sectors (Aerni et al. 2011).

As for the WTO Agreement on Agriculture (AoA), many legal scholars agree that the WTO did not reduce agricultural protection and subsidies but actually legalized them (Desta 2002). Nevertheless, Rodrik uses Isaiah Berlin's essay on the fox and the hedgehog ("a fox knows many things, but a hedgehog one important thing") to portray the WTO as a product of the hedgehog mentality of economists who are just concerned with the one thing they know, namely calculating efficiency gains from trade. He distances himself from them by regarding himself as an economist with a fox mentality that knows that context matters a lot in economic policy. Yet, if context would indeed not matter in the WTO, countries would have little interest in joining the club. All WTO Agreements contain an expansive set of policy options that is hardly made use of (especially when considering the wide gap between bound and applied tariff rates) (Aerni et al. 2011). Even in the case of intellectual property (IP) rights, the highly vilified TRIPS Agreement has probably made it easier in many respects rather than more difficult for WTO members to obtain access to important patented technologies. The 'compulsory licensing' amendment certainly helped to improve access to essential medicine in developing countries (Watal 2014).

So if Rodrik wishes to regard himself as a 'fox,' who draws on a wide variety of experiences and for whom the world cannot be boiled down to a single idea, he should also acknowledge the fact that the WTO was not an ideological but a pragmatic initiative. He rightly argues that it is the diversity of its domestic economic activities rather than the participation of a national economy in open markets that enables a country to embark on a path of sustainable growth. However, a national economy that attracts foreign direct investment (FDI) is also more likely to diversify its economic activities. After all, MNEs that decide to do business in a country may not just bring new physical goods into the domestic economy, but also introduce valuable knowledge on how to produce these physical goods and commercialize them effectively. This new useful knowledge is embodied in protocols, designs, and instructions. In this form, knowledge is a non-rival and partially excludable good and therefore has the big advantage of not being a scarce resource. In fact, the more the particular knowledge is used, copied and applied, the higher its value. Alas, trade statistics do not show the welfare effects generated through the exchange of these non-material goods (Jones and Romer 2010). China's creation of special economic zones, its policy focus on attracting knowledge-intensive FDI, as well as its huge investments in raising the human capital stock especially in the fields of natural sciences and engineering, are crucial pillars of progressive facilitating policies designed to enable rather than merely regulate economic and technological change (Ren et al. 2015). This also explains why China's development contributed most to the reduction of poverty during the period of the so-called Millennium Development Goals (MDGs) from 2000–2015. A UNDP report on China (UNDP 2015) attributes the country's success to three related factors: rapid economic expansion coupled with gradual reforms; policies that focus on inclusive growth; and demographic changes. Foreign aid or protectionist policies proved irrelevant.

Finally, Rodrik's recent attempt to defend the scientific value of the principles of comparative-static neoclassical economics, as long as these principles are not applied in a dogmatic or prescriptive way (Rodrik 2015), stands in contrast to his description of the economic success story of China. Only through a dynamic understanding of economic development can China's facilitating policies, which eventually enabled its economic rise, be understood. Such an understanding is derived from the theory of the great critics of textbooks economics such as Joseph Schumpeter and Paul Romer. Rodrik largely ignores their work.

2.3 Polanyi as 'Intellectual Guide' in Economics and Anthropology

Many other contemporary left-wing neoclassical economists, such as Thomas Piketty, Joseph Stieglitz, Paul Krugman, Jeffrey Sachs and Heiner Flassbeck who criticize the mechanical thinking of their own guild essentially reaffirm Polanyi's historical analysis, often without citing him, or any other social scientists who cannot be found in the textbooks of neoclassical economics (Panitch 2015). At the same time, Polanyi's Great Transformation remains one of the most cited works in economic anthropology and has regained momentum in the field through the recent publication of David Graeber's 'Debt: the first 5000 years' (Graeber 2011) and the new Routledge edition of Marshal Sahlins 'Stone-Age Economics' (Sahlins 2013).

Polanyi's work is generally celebrated by the anti-globalization movement as a grand narrative against the commodification of labor and land (Bugra and Agartan 2007). It also provides intellectual guidance to many social movements and schools of political thought today such as De-Growth, Primitivism, Nationalism, Food Sovereignty and Eco-feminism (Martinez Alier 2012). What these schools have in common is the belief that globalization must be something external and unnatural that has been forced upon humanity by undemocratic means. In search of an alternative, something that is alternatingly associated with 'substantive,' 'natural,' 'self-sufficient' and 'embedded,' globalization's critics tend to praise the self-sufficient economic systems that, so they assume, must have existed prior to capitalism. The nostalgic view of the past then turns into a sort of utopia that can be achieved in the future if only people distance themselves from the consumption culture of capitalism, get rid of their 'false consciousness' and eventually regain their self-determination. It is seen as a sort of life after 'Neoliberalism'. In this context, Polanyi's Great Transformation serves on the political left as a sort of substitute for Marxism in the post-Cold War period. Even though Polanyi does not offer any grand theory of capitalism and how to overcome it through class struggle, he shares the almost religious mythical framework consisting of the original paradise (life before capitalism), the fall from paradise (age of capitalism) and the kingdom of heaven to come (the New Jerusalem) built upon the ruins of capitalism. Marx, as well as Polanyi, offer meaningful world-historical narrative based on the utopian belief in the inevitability or necessity of overcoming capitalism to reach a state where social inequality is eliminated and humanity and social justice restored. They both ignore, however, the quintessential bourgeois institution without which humanity would not reproduce, which is the family (Nathans 2017). Having a family changes priorities in life, especially, if you belong to the class of servants rather than masters. For the family of the servant, income and private property may matter more than for the master, who takes it for granted and is largely concerned with postmaterial issues such as prestige (Kojève 1975). The primary purpose of the servant in his efforts to acquire material goods is to ensure a better life for his offspring. Therefore, the motivation to become an entrepreneur may bear no relation to a selfish and rational utility-maximizing individual as represented by the homo economicus in neoclassical economics. It is much more about the passion to serve others (with something they are willing to pay for) in order to eventually have the means to serve the needs of once own family and community. The zeal to reach utopia or individual self-realization, therefore, decreases with parental responsibilities. The next generation may still be concerned with self-realization as long as they have no offspring but the utopian views of a perfect society free of inequality and social injustice (including the utopia of a perfectly efficient society in a pareto-optimal state, governed by the principles of neoclassical economics) cease to matter. After all, the offspring was not involved in the 'great transformation' experienced by their parents, but are merely asked to memorize it from schoolbooks (Slezkine 2017).

Chapter 3
Neoliberalism: A Mythical and Meaningful Term Devoid of Any Deep Thought

Social scientists who write in praise of Polanyi, or reaffirm his insights without citing him, contribute to the construction of a powerful myth surrounding the term 'Neoliberalism' in contemporary discussions of globalization. In this mythological narrative, Neoliberalism stands for a wide range of undesirable trends, such as growing inequality, injustice, deprivation, fiscal austerity, deregulation, privatization, land grabbing, mental health problems, environmental destruction, and so on. As such, it is implicitly understood as an external force that has to be pushed back to restore the original state before 'Neoliberalism', which is assumed to be characterized by fairness, equality, solidarity, self-determination, freedom and environmental sustainability.

The construction of the myth of the golden past that must be restored erases all alternative interpretations of history and makes people lose the curiosity necessary to better understand how things actually evolved. In order words, the myth is becoming a natural fact that no longer needs explanation. As such, the belief that 'Neoliberalism' rules the world cannot be challenged and is undeniable since it contains an essentialist understanding of history that is devoid of contradictions but also devoid of any deep thought (Barthes 2012: 296).

When applied to the political discourse, any actor involved in business activities with a global reach is assumed to be a beneficiary of Neoliberalism and implicitly condemned for being motivated merely by profit to accumulate material goods. This assumed greedy attitude is then made accountable for the global social and environmental problems. Those employed by an MNE (as a signifier of neoliberalism) may object to the assumption that it would only be material interests that drive them. They may argue, for example, that their job is fulfilling and challenging, that they believe in the service they deliver to society or that their employment supports family members and relatives in need. These non-material and personal motives do however not matter to those who oppose the MNE that employed them. The account about their alleged non-material motives is discounted as an attempt to divert attention from the real corporate motives as described in essentialist psychology and anti-neoliberal literature. The same applies to groups identified as the

P. Aerni, *Global Business in Local Culture*, SpringerBriefs in Economics,
https://doi.org/10.1007/978-3-030-03798-7_3

victims of the corporate world such as those representing vulnerable cultural minorities. Many individuals within these cultural minorities may not want to be passive victims. They may see room to benefit from selective participation in economic globalization. Since their view does not fit the mythical bipolar account, their stories fail to be taken seriously, as will be illustrated by means of concrete examples in Chap. 4. In this sense, mythical accounts about the presumed perpetrators as well as the presumed victims tend to erase the complex context-based stories of the people who know that the world is not black and white but also full of grey zones where informed decisions are always based on compromise.

Even though the semiologist Barthes (2012) mainly analyzed the myths underpinning the belief systems of the dominating right-wing and bourgeois ideologies of his time in France, he also emphasized that myth construction in politics is not limited to one side of the political spectrum. According to Barthes, we are all at risk of being deprived of our language and history and defined by the timeless motives or stereotypes that our accusers attribute to us. In this context, the lazy explanation of parents to the 'why' questions of their children, namely that they should simply accept things the way they are, is symptomatic for the smug bourgeois society as well as the ideologues that claim to fight against 'Neoliberalism'. Both have ceased to be critical citizens because they have embraced popular myths as if they represent an objective truth. A political discourse that wants to overcome the shallowness of mythical accounts must be prepared to question the prevailing contemporary narrative—no matter if it is shaped by right-wing or left-wing populism. Such questioning must start with an inquiry of the past.

3.1 Did the Rent-Seeking Economy of Feudalism Serve the Needs of the People?

The economic system prior to industrialization in Europe was characterized by various types of feudalism. The foundation of European feudalism was the medieval manorial system, which was in turn derived from the Oikos system in ancient Greece and the Domus system in ancient Rome (Tauger 2011). In these self-sustaining economic systems, it was the property-owning Lord of the Manor, or Pater familias, who was vested with divine legitimacy to rule over his subjects (family, dependents, and slaves). The primary purpose of his rule was to ensure domestic self-sufficiency through economic autarky. All economic activities were primarily designed to meet the essential needs for shelter, clothing, and food of the manorial community. Even though the manorial system provided for common property to be used by the subjects, it was embedded in a static hierarchical patron-client system that did not allow for any social mobility. Moreover, it imposed a heavy burden on sharecropping farmers who were obliged to hand over a significant share of their harvest to the landlord as payment in kind, in return for protection (Braudel 1982).

According to Polanyi, the manorial system was, however, a more 'natural' economic system because it was firmly embedded in local tradition and culture, and based on reciprocity, redistribution, and autarky (Polanyi 1944/1978: 45–58). This would have ensured the right of peasants to use common property and to enjoy protection against military invasion.

3.2 The Enclosure Movement in the UK as the Beginning of Industrial Agriculture

In the 17th century, the right to common property was on the retreat in England with the establishment of the constitutional monarchy, and the subsequent emergence of a complex state bureaucracy that was run by the landed gentry. It pushed for investments in modern agriculture in response to the growing demand for food and animal feed stemming from the expansion of the urban population. This development led to the acceleration of the so-called enclosure movement. It brought about structural change in the countryside by enclosing the commons to promote 'private property'-based commercial agricultural systems. With thc assistance of Dutch agricultural engineers, farm management and, with it, productivity of agriculture greatly improved (Fagan 2001). This rural transformation surely generated social disruption, especially with those farmers who were expropriated by force. According to Polanyi (1944: 35–44, 101, 143), it was only thanks to the royal intervention, and against the will of the landed gentry, that the negative social impact of this rural transformation was mitigated through legislation in support of the poor and the dislocated (e.g., Poor Law, Speenhamland Law). The repeal of these laws of social assistance in the 19th century would have paved the way for the unfettered 'self-regulating' economy that fueled industrialization in England and later continental Europe. The social costs resulting from the expansion of unregulated markets would eventually undermine national culture and sovereignty, fuel nationalism and end in World War I, according to Polanyi.

What Polanyi failed to mention, however, is that the primary drivers of globalization in the 19th century were not just causing external social costs but also substantial external social benefits. In particular, the numerous public-private partnerships that were created to cope with the challenging social and environmental problems of industrialization were benefiting the poor in particular by contributing to improved hygiene and sanitation but also access to basic services in the areas of health care and education. This again contributed to greater social mobility and the creation of an entrepreneurial middle class (Morris 2000). In this context, the period could also be called the empowerment of the servant from the master (Kojève 1975).

Moreover, enclosed commercial estates in England that adopted many innovative crop-rotation and drainage techniques from the Netherlands played a crucial role in coping with the very low average temperatures in Europe in the late phase of the little ice age in the 17th century. Cold conditions decreased yields of traditional grains substantially and increased the threat of hunger and starvation (Fagan 2001). The enclosure movement also enabled structural change in agriculture that eventually helped to meet the growing national demand for food and animal feed due to population growth, urbanisation, and increasing affluence. In this context, research showed that European countries that continued to rely on feudal estates with common property, such as in France prior to the Revolution in 1789, were more affected by hunger and malnutrition than countries that promoted the expansion of commercial agriculture, such as England and the Netherlands (Walter and Schofield 1991).

Finally, the enclosure movement also resulted in increasing rural-urban commercial linkages that helped to transform the rural economies in a way that diversified local economic activities and increased average rural incomes (Fagan 2001; Braudel 1982).

3.3 Blaming Agricultural Trade Has Never Solved Any Food Security Problem

These empirical insights stand in sharp contrast to Polanyi's assertion that the liberalized grain market has been the cause of famine during the age of transition towards commercial agriculture (Polanyi 1944/1978: 256). He may have ignored the fact that the people most vulnerable to hunger and malnutrition back then as well as today in low-income countries are the inhabitants of marginal rural areas who are not connected to formal markets. Why? Because whenever a surplus is produced in the remote farming community, it is likely to go to waste because there is no formal and transparent market to sell it for a decent price; and whenever harvests fail, lack of purchasing power and high transaction costs discourage farmers in surplus regions to sell their produce in the famine-affected region (Aerni 2011a).

Since Polanyi follows Aristotle in his normative view that agricultural production should be designed for local use, not for trade and personal gain (Polanyi 1944: 56), he fails to recognize the potentially positive societal side effects of an entrepreneurial agricultural system. For example, the introduction of the land grant college system after the civil war in the United States in the mid-19th century was designed to bring innovation and entrepreneurship to impoverished people in the midwest of the country through state-supported institutions of higher education with a focus on the creation and dissemination of applied research in agriculture (Juma 2011; Aerni 2007a). The main purpose of these community-focused agricultural colleges was to address economic challenges and realize economic opportunities effectively. This decision to mobilize knowledge and technologies for regional economic development rather than merely provide subsidies to support vulnerable livelihoods proved to be highly successful in enabling economic empowerment and endogenous growth (Aerni 2011a; Etzkowitz 2004; Etzkowitz and Zhou 2017).

A similar example is New Zealand in the late 20th, when the government liberalized the national agricultural system in the 1980s with the support of the federated farmers of New Zealand (Aerni 2009). In Rodrik's understanding of the globalization paradox, the decision of the federated farmers would make no sense because farmers are supposed to be the victims of economic globalization and therefore ask for protectionism and the reassertion of national sovereignty, rather than endorse economic liberalization. Yet, the farmers in New Zealand did not want their government to continue to act as a nanny nurturing a grievance culture in rural areas with ever more subsidies per capita. Instead, they preferred the government to become a coach, assisting them with venture capital and marketing, while incentivizing the national agricultural research system (Crown Research Institutes) to work more closely with entrepreneurial farmers and the private sector in the development of new products and services that help diversify national agriculture (Aerni 2009).

As a robust democracy, New Zealand is also one of the countries that contradicts Polanyi's view that market-oriented societies evolve toward fascism (Cornell et al. 2017); and it is a great example to illustrate that the best way to empower women is to give them economic rights before political rights. In 1893, New Zealand became the first country in the world in which all women had the right to vote in parliamentary elections. The reason was the prior economic integration of the female workforce. Women were desperately needed in this remote country to build up the economy. Based on their economic importance, the increase in political influence was only logical.

3.4 Fernand Braudel's Criticism of Polanyi's Interpretation of History

Polanyi may have overlooked the oppressive and exploitative character of the former feudalist self-sufficient economic systems because his archival research had a clear selection bias. He especially ignored the archival material of the French École des Annales, of which Fernand Braudel was the most prominent representative. As a social historian, Braudel illustrated the misery of the sharecropping farmers in the embedded traditional rent-seeking economies of feudalism. For him, the feudalist system was less about society subordinating the economy than about the landlord controlling and exploiting his subjects by extracting ever higher rents from the fruit of their labor (Braudel 1982). In this context, Braudel found the large feudal estates in Eastern Europe of particular cruelty and unfairness. The landlords of this region were eager to integrate their traditional economic system into the emerging capitalist system, yet instead of investing in entrepreneurship and innovation they preferred to squeeze more back-breaking work out of their farming subjects.

As a consequence, indentured peasants had no incentive to invest in the improvement of the quality and quantity of their agricultural produce. After all, agricultural surpluses and added value of food products primarily benefited the cruel rent-seeking landlord—not the peasants and their families. To escape the oppressive system, many peasant families migrated overseas in the hope of finding more freedom and better economic opportunities elsewhere.

Even farming families in Swiss mountain villages who tended to be less oppressed and enjoyed considerable autarky, were forced to migrate, because population growth combined with common property management has led to the overuse of natural resources. This migration from marginal areas largely accounts for the fact that Switzerland became the most successful exporter of professional mercenaries in Europe (Holenstein et al. 2008). The main reason for the large emigration waves of Swiss people toward overseas in the 19th century was the ever-shrinking average farm size. Because of population growth and lack of off-farm employment, farm households were forced to subdivide their already small piece of land among the numerous offspring ('Realteilung'). Eventually, the size of the plot per family member no longer sufficed to feed the next generation. This led either to logging or emigration across the countryside of Europe to cities and abroad (Sauerländer 2016). A similar trend can be observed today in many African rural economies with high population growth and lack of off-farm employment (Aerni 2015a). However, in today's debate about the root causes of migration in Africa, the result of a continuously shrinking average farm size, has so far been ignored in the political debate, because history has been largely converted into mythical storytelling in the Polanyi style—and shrinking farm sizes do not fit the popular mythical account small-scale farmers feeding the world with their traditional agricultural practices.[1]

Braudel who reconstructed the precarious life of ordinary people in feudalist regimes complained about sociologists, economists, and anthropologists who 'have unfortunately accustomed us to their almost total indifference to history' (Braudel 1982: 227). Braudel (1982) and Weber (1992) also rebutted Polanyi's historical claim that formal market exchange, which includes trade in what Polanyi calls 'fictitious commodities' such as land, labor, and money, would have been absent in traditional and non-Western societies.

3.5 Why Polanyi's Bipolar Framing Finds Fertile Ground

Despite its contradictions, Polanyi's narrative of society versus the market-based economy has become very popular with people in affluent Western economies who loathe the risks and uncertainties of global economic and technological change.

[1]The report of the International Assessment of Agricultural Knowledge, Science and Technology for Development (IAASTD 2008) was crucial in creating this mythical account. See report https://www.globalagriculture.org/report-topics/about-the-iaastd-report.html (Accessed Sept 25 2018).

It provides a meaningful framework of interpretation for undesirable societal trends ranging from the precarious and uncertain living conditions of the youth, social inequality, the depreciation of labor, as well as widespread physical and psychological stress symptoms presumably resulting from economic pressure (Furedi 2003).

No matter how compelling Polanyi's diagnosis of modern society's ailments and his suggested remedies may sound, the account suffers from several inconsistencies. First of all, it lacks a comparative dimension assuming that all other possible economic systems have never been tried before. History is littered with visions of utopia that eventually turned into dystopia once put into concrete practice. The major flaw of these visions was the failure to understand the proper nature of the human being. They assumed that rules based on fairness and reciprocity that work well with small, homogenous and informal communities, may also apply to large, formal and heterogeneous societies. This 'ecological fallacy' ignores the fact that we all have to learn to live in two worlds. On the one hand, the anonymous world of economic exchange in which the individual must learn to realize his or her potential within the boundaries set by the formal rules of the economic and political system; on the other hand, there is the personal world of social exchange. It is based on the informal rules of fairness and reciprocity within one's own family and community (Hayek 1944; Smith 2003; Ariely 2009). The necessity to learn to live in two worlds is not a new phenomenon but is part of the human condition.

Large anonymous societies function because the formal rules that govern anonymous market-, as well as political transactions, are based on the assumption that people pursue their self-interest. They do so, not necessarily because they only care about themselves, but because they have to care for their dependents too. In other words, democratic constitutions and the rules of the formal market economy share a similar understanding of human nature and, as such, can be considered intellectual twins (Hirschman 1992).

Another flaw of Polanyi's narrative is that it starts from the general public resentment and personal grief about global economic change and then identifies a scapegoat in the form of 'the system' and 'its agents' (e.g., economic globalization and the MNEs that drive it). Facts are then only taken into account if they confirm the role of the assumed victim and the role of the assumed perpetrator (Nickerson 1998). This confirmation bias has severe consequences when it is built into the public discourse. It leads to an increasingly defensive world view manifesting itself in the resentment against the growing uncertainty and complexity of globalization and the longing to return to a more simple past consisting of predictable and manageable traditional local practices and habits (Schumpeter 1942).

Many scholars argue that the bipolar mindset of the 'good' local versus the 'bad' global was shaped during the period of the Cold War. At that time, it was related to the permanent threat of mutual annihilation between the two superpowers, the United States and the Soviet Union (Hamblin 2013). Such a mindset of 'we versus them' may have been meaningful in the past geopolitical environment; it is, however, inadequate to deal with the opportunities and challenges of the global knowledge economy with its shrinking geographical distances and the potential for economic and technological leapfrogging. Yet, the bipolar mindset still prevails in many areas of academia and civil society today. The watertight distinction between the local and the global has affected academic research in the social sciences in particular. Influential professors who were trained during the Cold War period tended to build their research upon theories of the 1970s that essentially argued that poverty in one region is the result of affluence in another region. This belief was passed on to the next generation. The reaffirmation of these theories became a selection criterion for future faculty appointments (Lässig 2016; Bourdieu 1988).

Noble-prize winning economist and social activist Joseph Stiglitz represents to some extent the academic who successfully applies the bipolar mindset of the Cold War to the challenges of economic globalization in the 21st century by building upon Polanyi's essential distinction between the embedded local economic system and the dis-embedded global economic system. In his foreword to the 2001 edition of 'The great transformation', he argues that economic globalization tends to result in job destruction rather than job creation in low-income countries. With this claim, he essentially embraces the traditional belief that formal markets expand at the expense of embedded traditional markets. The nostalgic view is contradicted by the most significant achievement in poverty reduction over the past three decades, which is the economic rise of Asia and its integration into the formal global economy (Khan 2007; UNDP 2015).

Chapter 4
The Impact of Popular Stereotypes in Academic Research and Public Policy

Schumpeter (1942), and most lately, Cowen (2017) have observed a trend toward bipolar thinking in affluent societies that do not need to bother anymore about material necessities. It is based on the essential distinction between 'us' (the value-based community) and 'them' (the value-less agents of global change in pursuit of material benefit). It also manifests itself in the perception of economic and technological change as a threat to personal and collective identity and integrity. Since rootless and profit-seeking MNEs are seen as the main drivers of global change, they are resented by identity movements on both sides of the political spectrum. They are accused of dis-embedding the local, self-sufficient economy, of endangering food sovereignty and cultural identity, of lack of patriotism, and of destroying the natural environment. In this sense, Polanyi's narrative appeals to progressive as well as reactionary circles because of the fear of environmental and cultural contamination of the idealized local conditions, portrayed as pure and authentic (Uekoetter 2006). The fear of contamination manifests itself also in the 'Not In My Backyard' (NIMBY) attitude among privileged insiders with a strong sense of entitlement and a focus on preserving things rather than changing things. The privileged insiders that form NIMBY communities consider themselves progressive and may voice their concerns about the victims of global change. However, in their daily lives, these so-called hipsters are mostly concerned with beautifying, embedding and preserving their value-based but increasingly segregated neighbourhoods. They are tolerant towards underprivileged outsiders and think that they should have a right to be anywhere except in their backyard (Current and Tillotson 2018).

P. Aerni, *Global Business in Local Culture*, SpringerBriefs in Economics,
https://doi.org/10.1007/978-3-030-03798-7_4

4.1 Echo Chambers: The Attack on Democracy from Within

Joseph Schumpeter explained the growing hostility of affluent societies toward innovation and entrepreneurship with the fact that capitalism can provide human beings with everything except meaning, identity, and orientation (Schumpeter 1942). Enlightened individuals are expected to search and identify these non-material needs on their own. However, since this is frustrating and time-consuming, many markets for meaning and identity have been created in postmaterialist, service-oriented affluent societies by offering mythical accounts that are simple to understand, sound intuitively right and feel meaningful (Aerni 2011b). In this market of postmaterial values, political entrepreneurs as well as spiritual leaders provide a narrative for external political as well as individual internal crises and offer a way out by refocusing on alleged lost cultural values and norms as well as the pure inner authentic self. Such narratives reduce complexity by providing a simple frame for interpreting political events as well as for guidance in daily individual life (Luhmann 1993).

As a consequence, postmaterialist societies tend to replace kinship-based communities with value-based communities that are often more united by a common enemy than shared values and norms (Greene 2014). The result is often less interaction with people beyond the like-minded community, online and offline (Sunstein 2007), as well as a tendency to regard one's personal opinion as something authentic that represents part of one's identity (Aerni 2011b). As a result, changing an opinion in response to new insights or a convincing argumentation by someone else almost always amounts to a feeling of betrayal toward oneself. However, a lively and functional democracy relies on people learning to live in two worlds, the private realm of social exchange and the public realm economic exchange. In this context, the public realm of economic exchange cannot work by merely voicing private views in public. After all, the main purpose of exchange in business and politics is to reach a deal or a political compromise, not to merely agree to disagree (Van Winden 2007).

So ultimately, democracy may not face its biggest challenge from without, in the form of globalization and authoritarian regimes in the developing world, but from within, in the form of complacent postmaterialist societies in affluent economies that prefer to stick to their homogeneous and segregated communities rather than engage with other parties that represent different views, experiences and interests (Tyler 2017). Richard Sennett anticipated this trend in the 1970s in this book 'The Fall of Public Man' (1977). More recently, Cass Sunstein warned that the internet, and especially the social media revolution, is likely to accelerate the trend in the 21st century through echo chambers that enable individuals to filter out any new online information that would not be in line with personal prejudice and thus produce unwanted cognitive dissonance (Sunstein 2007).

Such value-based homogenous communities may feel firmly embedded in a cozy and conflict-free social network of like-minded people. But they do not realize that their lifestyle, even when frugal and mostly focused on local consumption, is highly dependent on technology, stakeholders, and institutions that have no local roots.

The exclusive framing of economic globalization as a threat to shared community norms and values has become a global phenomenon. It risks creating a monoculture of the self-righteous mind flourishing in a particular milieu that defines itself as 'we' against 'them', a sort of generic 'amoral' global anonymous society (Haidt 2012). Whoever does not agree with their views must be an agent of 'them', brainwashed by marketing or merely naïve. This alarming trend toward self-righteousness and quasi-religious views does not just manifest itself in the public discourse on politically controversial topics related to health, agriculture and the environment, but also in debates held at universities. Universities used to be the safe harbor of freedom of speech and characterized by the contest for the better argument. However, student movements and academics in the field of the social sciences especially at elite universities in prosperous Western economies, have increasingly promoted self-censorship regarding politically sensitive topics that are potentially offensive to the values and identity of particular students, who may then need counselling (Furedi 2016). In this context, academic freedom may conflict with academic policies concerned with damage control, wellness, and reputation (Williams 2016). Especially in the social sciences, the focus of research and teaching is moving toward topics that are assumed to be popular with students and thus potentially reputation-enhancing such as gender, ecology, organic agriculture, food sovereignty and human rights. These fields of research and teaching are often touted as interdisciplinary, ethics-oriented and as offering critical viewpoints. However, a closer look often reveals that they share somewhat conformist baseline assumptions about presumed perpetrators and victims in the age of globalization (Williams 2016). The implicit view is that economic and technological change is excellent as long as it does not harm anyone, emotionally or physically. However, since any change does not just create opportunities but also risks, harm is likely to happen at some time, somewhere, no matter how many precautionary measures are in place; as such 'harm caused by change' represents an exhaustible source of outrage and, in response to it, available research funding to address the problem.

4.2 Embedded Liberalism: A Flawed Concept

John Ruggie, an influential political scientist, CSR scholar and main contributor to the UN Guiding Principles on Business and Human Rights (UNGP) (Ruggie 2008), is seen as one of the fathers of 'embedded liberalism', a school of thought in political science that mostly builds upon the line of argumentation of Karl Polanyi and chimes well with the post-material moral values of Western, educated, industrialized, rich and democratic (WEIRD) societies (Haidt 2012). By implicitly framing global business as a threat to local human rights, Ruggie very much appeals to the popular view that FDI must be tamed and regulated to protect human rights, especially in developing countries with weak institutions that prove unable to enforce strict labor rights and environmental regulations.

Embedded liberalism, as advocated not just by political scientists such as John Ruggie but also economists such as Dani Rodrik, also suggests that governments should not open certain domestic markets to global competition unless they have a strategy on how to compensate the potential losers through the creation of a generous social welfare state. Whatever the social benefits that resulted from the creation of the welfare state in Europe and North America during the Cold War period, it certainly did not lead to more but probably less embeddedness. The disembedding effect is related to the fact that the welfare state freed the individual from dependence on his or her kin and local community and created a sense of individual entitlement but also vulnerability (Romer 1996; Luhmann 1993).

In a sense, the narrative of 'embedded liberalism' provides incumbents in local business and politics with an opportunity to portray themselves as victims of the expansion of an increasingly disembedded global economy. They can be confident in gaining support from the numerous NIMBY communities which also share the need to defend the local against the global. The joint demand for public support and regulation against allegedly predatory MNEs is therefore likely to resonate with local policymakers serving the agenda of the entrenched private interests that control the local economy wrapped into a language of moral concern (Buchanan and Tullock 1962; Aerni 2011b). In the United States, the rhetoric of Donald Trump is very much in line with the view that rootless MNEs may damage the local economy. His displeasure with their lobbyists in Washington DC who seek political support for an open trading system may explain his broad public support.

4.2.1 The Artificial Separation of the 'Authentic' Local from the 'Generic' Global Product

Ultimately, it is perception, not facts, that matter (Slovic and Lichtenstein 2006). Local people prefer the familiar, the local store selling local products and services as they represent part of local identity. There is nothing wrong in favouring the local variety, but it should be acknowledged that local products and services have ceased to be de-coupled from the global economy, especially in high-income countries. Even in the highly subsidized rural areas of these wealthy countries, the local grocery shops are increasingly run by migrants. Migrant workers are also mostly accountable for the construction of the buildings, the harvesting of the local food, and for the caring of the elderly.

When it comes to agriculture, local food production is highly dependent on imports, such as greenhouses, post harvest technologies, plant protection, seeds, and fertilizer (Freidberg 2009). For marketing purposes, local products are nevertheless praised as real alternatives to the global dis-embedded economy.

This marketing concept is successful because of the public distrust directed against agents of global economic and technological change who are seen exclusively as a threat to the business model of local incumbents (Aerni and Bernauer 2006). The drivers of this change may not even be large MNEs but actually small innovative start-up companies that are able to offer better and more innovative services and products to local people than local competitors. In view of their more scarce resources, these small firms may suffer more from the costly regulation resulting from the political rhetoric against MNEs than MNEs themselves. Acting as a protector of local interests, a NIMBY community may preserve its 'embedded' economy, but it bears little resemblance to Polanyi's understanding of an embedded economy based on fairness, reciprocity, redistribution, and autarky. Moreover, it may not be true that global economic integration leads to the disappearance of local traditional practices and techniques. Jane Jacobs convincingly argued that the probability that old goods and services are finding a new commercial use, is higher in societies that are more open to technological and economic change. She calls it the 'backward' application of new techniques to goods and services that would otherwise go extinct (Jacobs 1970).

4.2.2 Governments as the Blameless Defenders Against Careless Big Business

In defense of the principles of embedded liberalism, Abdelal and Ruggie (2009) argue that governments need to promote a corporate culture respectful of human rights and environmental sustainability at home and abroad. They should do so through pro-active regulation designed to minimize risks. As such, corporations increasingly assume the guardian morality of governments in the hope to be seen as defenders of the moral order. Yet, it is not surprising that the public distrusts such claims because the purpose of corporations is to generate value for society by taking risks, not by avoiding them. As a consequence, voluntary actions and codes of conduct developed and implemented by MNEs to minimize the social and environmental costs of their business activities is often interpreted in the public debate on business and human rights as mere fig leaves designed to disguise their real motives, namely to expand their global business at the expense of local firms.

In view of the failure of voluntary action to induce more trust in corporate activities, Abdelal and Ruggie (2009) suggest that they need to be complemented by legally binding regulations to protect human rights from harmful corporate practices in countries with weak enforcement capacities. They seem to ignore that plenty of regulation is already in place to discourage illegal, harmful corporate practices, and that monitoring and detecting fraud has become easier with technological change. A combination of modern biotechnology, nanotechnology, and digital technologies makes it possible today to detect harmful substances at the molecular level. This is creating a lot of fearful media alerts even though most substances are only harmful when they exceed a certain threshold of quantity

(Kovács et al. 2014). Moreover, public shaming in social media has also become an additional powerful tool to discourage unsustainable business practices, even if the practices may not be strictly illegal or if the allegations prove to be wrong. Negative media coverage, whether justified or not, can have a serious material impact on the targeted company, not just because it harms the company's image with clients, consumers, and rating agencies, but also because adverse media coverage makes recruiting talents more difficult (Petley 2013).

4.2.3 Why More Regulation Does not Lead to More Public Trust: The Case of GMOs

Ruggie's argument that public trust in a particular industry can be regained by regulating it more strictly, does not seem to hold in practice (Harring 2018). In view of concerns about health and the environment, public trust in the way food is produced is crucial and explains why there is so much regulation and self-regulation in the business of food and agriculture. However, did all this regulation increase trust in companies that breed and grow crops, raise cattle and process and commercialize food? The experience with preventive regulation of genetically modified organisms (GMOs) in food and agriculture in Europe indicates that this is not the case. Governments in Europe embraced the risk-based media discourse on GM crops at an early stage. In the absence of any scientific evidence about genuinely new risks to be attributed to genetic engineering, governments justified lengthy and burdensome regulatory approval processes of GM crops and mandatory labeling for GM food by arguing that this would help regain trust in the new technology and the companies that use it. Yet, the effect has been the opposite. In view of all the new risk regulation, people felt confirmed in their assumption that there must be something fishy with this technology and its related business (Aerni and Bernauer 2006). Apart from not contributing anything to public trust, costly regulation increased industrial concentration in agribusiness, because small innovative companies could no more afford to comply with lengthy and burdensome applications for field trials and commercial releases (Bonny 2017; Smart et al. 2107). Besides, the organic farming business in Europe has mostly moved from being a small and innovative community that offers alternative food products, to a large commercial business promoted by global retailers and with the support of governments. The organic industry defines itself mainly as being free of genetically modified ingredients. Its active anti-GMO marketing must be understood as an effort to secure trust with taxpayers and consumers as providers of local, authentic, healthy and natural food as opposed to the food provided by big corporations. The resulting political polarization made it impossible so far to combine agro-ecological approaches with the advanced tools of modern biotechnology in efforts to design context-based tailor-made solutions. The consequences for the sustainability and the affordability of ecologically produced nutritious food were clearly negative (Aerni 2015c; Ronald and Adamchak 2008).

4.2.4 Self-Regulation in Industry as a Base for Subsequent Government Regulation

Based on the assumption that MNEs investing in local economies are likely to generate local social and environmental costs, Ruggie developed the "Protect, Respect and Remedy" concept, also known as the Ruggie Framework. It underpins the UN Guiding Principles on Business and Human Rights (UNGP), which is based on the principle of 'do no harm'.

The UNGP ignore that MNEs may have a long-term self-interest to treat employees with respect, to operate in an environment that is free of hazards to human health and to ensure consumer satisfaction. The history of industrialization in Europe illustrates this well. The first wave of globalization in Europe in the 19th century was accompanied by efforts of industrial associations to introduce self-regulation even before there was any government regulation regarding the protection of consumer interests (Ping 2011). Such efforts were based on self-interest because, if only one company were irresponsible, the whole industry would suffer from public distrust. As is the case today, it was all about the license to operate problem; after all, economic success is ultimately underpinned by social legitimacy (Deephouse et al. 2017).

It may be an indication that industry has always been aware of the fact that, in the absence of government regulation, self-regulation is necessary to discourage cheating and to ensure the respect and protection of stakeholder interests within the society they operate. The fact that business did not wait until government has passed regulations may also indicate that companies with a long-term perspective will act in a morally responsible way—out of self-interest. In this sense, MNEs that are committed to complying with industry-wide standards and are prepared to embed themselves into the local economy may actually help improve human rights conditions, as will be illustrated by using concrete case studies in Sect. 7.4.

4.2.5 Why the Ruggie Framework May Not Be Harmless

Ruggie's guiding framework for the UNGP with its defensive character (respect, protect, remedy) does not acknowledge any possible positive contribution of business to the improvement of human rights by defining corporate responsibility basically in negative terms, namely "to do no harm". This may sound fair, but it forces MNEs to focus their efforts exclusively on avoiding risks rather than taking risks to create not just economic but also societal and environmental value through scalable innovation (Aerni 2015b). After all, such innovation is crucial for economic empowerment through the creation of new markets with increasing returns. This usually also results in increasing incomes in the respective region, which automatically improves local access to essential human rights, such as the right to food, water, shelter, gender equality, etc. In addition, scalable solutions to sustainability problems tend to be developed by actors in industry rather than government or civil society. After all it is them who make use of new knowledge to develop substitutes for existing polluting products and processes and improve their resource efficiency (Naam 2013). The UN SDGs explicitly acknowledge the contribution of an innovative private sector to inclusive and sustainable growth (UNDP 2017; UNCTAD 2017b,c).

Ruggie's focus on risk avoidance may be especially counterproductive in low-income countries that desperately need economic and technological change to create a future for its well-educated but poorly-connected young people that live in the hope of applying their skills in business. Becoming a supplier of, or being employed by, a subsidiary of an MNE that operates in the country is not just well-remunerated but comes with capacity development and technology transfer. Moreover, the integration into global networks through contracts with MNEs increases the value of local business networks and, on the individual level, the chances of a good marriage and a better life for the offspring (Aerni 2015a). All these aspects are ignored by the UNGP, and therefore the focus on 'do no harm' may cause indirect harm by setting the wrong priorities. More embedded FDI may ultimately be more effective in providing improved material access to basic human rights than mere human rights advocacy work, even if such foreign investments do not come without risk.

4.3 Indigenous Communities as Projection Screens for Preserved Cultural Embeddedness

The cultural theorist and old-school public intellectual Terry Eagleton published a very entertaining book on 'culture' (Eagleton 2016). He describes the term in its narrow and broad conceptions and identifies 'tending to natural growth' as the actual root of the word 'culture'. After all, the word 'agriculture' indicates that the main task of mankind is to convert nature into culture in order to survive. This also constitutes one of the basic insights of philosophical anthropology (Carroll 2017). In other words, culture is less related to the conservation of cultural practices than the transformation of these practices through innovation. Eagleton would not go that far, but laments instead how the capitalist system has incorporated culture for its own material ends. This would have given birth to a global cultural industry that crushes local cultural practices that are still embedded in local language and local economic activities. He admits that a hybridization of the global and the local is not something new but warns about the consequences of hybrid cultures that comes with the global capitalist system. It would result in the longing for the authentic and the pure culture that must have existed before—which is at the root of right-wing (back to 'pure' national culture) as well as left-wing (back to 'pure' nature) populism.

Eagleton interprets these populist identity movements as a sort of natural self-defense against the cultural and economic hegemony of Western capitalism. He condemns 21st-century capitalism by using the arguments of the romantic thinkers around the end of the 18th century, such as the attack of Edmund Burke against British colonialism and Johann Gottlieb Herder's defense of ethnos defining people by their common language, cultural traditions, and myths of origin. Burke denounced the age of colonialism as a culture built on the exploitation of the ruled, disrespecting and disembedding the local culture it encroaches on. But is this comparable to the capitalist system of the 21st century, which is increasingly driven by non-Western emerging economies and cultures such as China? Unlike colonialism, 21st century capitalism appears to be quite attractive to non-Western economies. How is it possible to explain with Burke how capitalism actually enabled the economic and cultural empowerment of many non-Western societies? By not making a clear distinction between mere stealing, which was often the case with colonial powers invading foreign cultures, and formal rule-based trading and exchange between global and local economic actors, Eagleton follows classical Aristotelian thinking with its hostility towards markets. It is therefore not surprising that he eventually ends up with the same bipolar mindset of global business versus local culture that he very much criticizes with the postmodern culturalists, who would not make any normative distinctions between what is worth preserving and what is not. They would value diversity, so Eagleton argues, for its own sake and consider it their moral duty to save every possible form of cultural expression from alteration or extinction. However, this kind of belief in the authenticity and purity of culture and the support for cultural self-defense of vulnerable indigenous communities could have also been derived from Herder's philosophy.

Yet, the younger generations of indigenous communities prove increasingly unwilling to play the role of the passive cultural victim of a disembedding global economy, even though their opposition to it may endear them to affluent urban communities concerned with identity. Gaining the attention of these affluent people by approving of their postmaterial values may result in charity projects designed to support their endangered culture, but it limits their freedom of deciding on their own where to move with their culture and what to adopt from other cultures (Easterly 2002; Juma 2011). Many indigenous leaders realize that their culture cannot be kept artificially alive through subsidies and foreign aid. Moreover, such an approach fosters a grievance culture that stifles entrepreneurial initiative and discourages the young and educated from staying in the community (Porou 2013).

The emphasis of indigenous communities on respecting their economic rights (right to ownership, right to self-determination and economic development, equality before justice, freedom from discrimination) is most evident in Article 21 of the United Nations Declaration on the Rights of Indigenous Peoples from 2007[1]; it demands the right to the improvement of economic and social conditions through

[1]See: http://www.un.org/esa/socdev/unpfii/documents/DRIPS_en.pdf (Accessed Sept 25 2018).

access to education, vocational training, and capacity development. Moreover, Articles 23 and 32 underscore the right of indigenous communities to set their own priorities in development.

In particular, indigenous communities do not want to play the role of museum objects, because they have never been passive actors waiting for affluent visitors to be smiled upon. Especially the younger generation that witnessed the detrimental and disempowering effects of government-subsidized indigenous culture over the past decades is looking for cultural renewal rather than cultural preservation. Young people tend to be more experimental. They prefer to combine their traditional knowledge with modern and user-friendly practices and technologies. However, in order to earn a living from the creation of new markets, they must aquire the necessary business skills and have access to credit, infrastructure, and business services.

A concrete case study (discussed in further detail in Sect. 8.1.2) illustrates, for example, how a Maasai-owned business can become profitable for its local associates while simultaneously improve environmental management and ensure cultural self-determination (Aerni 2015a; Alarcon et al. 2017).

International advocacy groups, such as Survival International, that recruit many of their members from urban NIMBY communities (Eggen and Roland 2013) claim to represent the interests of indigenous people and their environment by defending their 'autarky' as traditional 'embedded' communities. Local communities may however disagree with this portrayal of their interests. In view of population growth and persistent poverty, they are unlikely or even unwilling to preserve all their traditional practices, unless they get remunerated for it. Receiving financial support from external donors to preserve traditional practices is however hardly sustainable because it does not increase but actually undermine the local autonomy of the indigenous community. Moreover, it is likely to accelerate rural-urban migration (Keller 2015; Aerni 2016).

Global activists who claim to defend the interests of indigenous people often tend to reframe the local story in a way that relates to global concerns because their constituency and their donors are located elsewhere (Rangan 2000). A struggle against the global forces also sells better to the global media than a struggle against unknown local actors, as is illustrated in the following section.

4.3.1 How the Indian Chipko Movement Became a Symbol of NIMBY Environmentalism

In her book 'Of Myth and Movements' (Rangan 2000) the Indian scholar Haripriya Rangan documents how global celebrities in the anti-globalization movement were able to hijack the cause of local protest in India for their own benefit. Rangan grew up in the Garhwal region at the slope of the Himalaya Mountains in the Indian state Uttar Pradesh. When she studied Geography in the United States in the 1980s, she learned about the roots of the popular tree-hugging movement, which could be traced back to the co-called Chipko Movement that emerged in her region in India during the 1970s. When doing her field research in her home region, she noticed that the movement was quite heterogeneous and the motives of resistance reflected local economic concerns that had very little in common with priorities of the global environmental movement. It was a resistance movement against the Forest Department of Uttar Pradesh (UP), which had assumed de facto control over the forested areas of the region and invested in afforestation projects undertaken to achieve self-sufficiency in industrial wood supplies. This government takeover over local resources led to a process of economic marginalization of small-scale extractors that was compounded by the state governments' failure to provide financial assistance for disaster relief during heavy monsoon rains in 1971 and 1972.

In 1973, the Forest Department of UP denied concessional allotments to artisanal cooperatives to make commercial use of their forestry resources. Instead, it favoured contractors from outside the state. Village leaders opposed this decision and urged their communities to prevent external merchants and contractors from extracting trees during the subsequent felling seasons. Men and women, regardless of age or caste, started to hug trees to prevent them from being cut and urged migrant labourers to return home. The Chipko protest coincided with the emergence of new social movements in the 1970s that were identity- rather than class-based; as well as the Stockholm Conference on the Human Environment in 1972, the first UN event dedicated to environmental concerns. Around the same time, the Indian Prime Minister Indira Gandhi promoted state-driven development aimed at creating full and fair employment and advancements in all regions of India via massive state intervention in various sectors of the economy. In this new socio-political environment, Sunderlal Bahaguna emerged as the charismatic spokesman of the Chipko movement on the national stage. He primarily adopted the anti-business rhetoric that resonated well not just with socialist politicians on the national level, but also environmental scholars and activists in other parts of the country. As a consequence, Bahaguna became a symbol of Indian Environmentalism, celebrated primarily by the Indian political elite who responded with national legislation aimed at discouraging private sector activities with forest products and promoting forest protection in the Himalayas.

In 1976 Indira Gandhi added Article 48A to the constitution "to protect and improve the environment and to safeguard the forests and wildlife of the country". The article gave the central government the power to overrule state decisions on environmental matters. These political and constitutional changes prepared the groundwork for the first national Forest Protection Act and the creation of a Federal Department of Environment in 1980. After having created a federal institution for Environment and Forestry and passing the first Forest Protection Act in the mid-1970s, the national government imposed a 15-year ban on the felling of green timber above 100 m in the Himalayas. In 1988, the Forest Conservation Act went even further and prohibited any individual or corporate institution from engaging in afforestation projects on state-owned land without permission from the Ministry of Environment and Forests (the formal Federal Department of Environment).

Environmental activism during and since the 1970s also had legal consequences at the state level. Uttar Pradesh (UP) passed the UP Forest Corporation Act in 1975 to control the wholesale supply for timber and other commercially valuable forest products. In 1976, the UP Tree Protection Act to protect valuable forests was enacted. The result was a massive increase in the administrative burden of forest management, including routine forest maintenance activities such as lopping, thinning and clearing. It also fostered the emergence of a 'timber mafia', many of whose members used to be legal forest contractors before losing their concessions. Not surprisingly, the village leaders who led the Chipko movement started to resent the unintended political consequences of this well-meant legal activism to protect their forests. Nearly two-thirds of the total land area in the region was classified eventually under different forest reserve categories. The use of forest resources in these areas required permission from the ministry, which tended to approve large-scale hydroelectric power projects swiftly but ignored small-scale development projects. The women of the Chipko movement complained that they were deprived of access to areas where they previously collected medicinal herbs and other commercially valuable forest products (Rangan 2000: 166).

Since these developments were never intended by the original resistance movement, the story of the Chipko movement is an example of how global and national trends dominate discursive strategies that ultimately affect the meaning of local activism. The interpretation of local events must fit the overall frame of the global narrative of environmentalism in order to be heard and raise political interest. The legal consequences of the distorted narrative are then felt again by the locals, who struggle for their economic rights but ultimately witness a systematic undermining of these rights allegedly for the sake of 'their' environment.

The tree-hugging movement, as one of the symbols of global environmentalism, ultimately reflected the romantic projections of urban elites in developed and developing countries about the 'sustainable' livelihoods of local communities without any real interest in understanding the actual concerns of those communities. The demands of the local men and women in the Garhwal region have been either ignored or completely misinterpreted.

In this sense, national and global environmentalists and development activists did not empower but disenfranchise the local community by portraying its people as representatives of a world that is detached from any commercial interests and in harmony with nature. Portraying them as gentle and wise but backward and therefore in need of protection, the wealthy elite made them a perfect symbol of purity and innocence and therefore vulnerability.

4.3.2 *Myths Embodied in Scientific Models that Guide Academic Research*

It would not be sufficient to merely attribute the collateral damage of global environmental activism on the economic rights of local communities to well-meaning affluent urban people that 'care' for the vulnerable communities and their equally vulnerable environment. After all, environmental scientists in academia, civil society and government argue that their actions are not based on populism but science-based findings from ecosystem research. This then serves as a means of justification for their legal actions. However, the ecosystem models that provide the theoretical underpinnings of this research are self-validating narratives by themselves. According to Rangan (2000), they are similar to myth in the sense that the abstraction does not correspond to reality, but actively distorts it instead. Ecosystems research tends to rely on the baseline assumption of an allegedly self-contained ecosystem as part of an internal self-regulating mechanism that ensures a return to equilibrium. This assumption has been criticized frequently, yet it continues to be used for ecosystem modelling because it makes intuitive sense and is easy to communicate (Maris et al. 2018). The methodology implies that everything is in harmony in such self-sufficient systems until they are disturbed by something 'external'. When the source of the disturbance has been identified and eventually removed, the original state of harmony may return. This popular metaphor is also applied in neoclassical economics where disruptive innovation is essentially treated as a 'black box' that causes business cycles that temporarily disturb the economic equilibrium until competition or government intervention of some sort restores or recalibrates it (Romer 1994; Aerni 2015a).

Last but not least, the sequential portrayal of events beginning with an original equilibrium and harmony threatened by an external force and ending with the restoration of the original state by neutralizing or expelling the external force is also implicitly part of most environmental narratives. As such, it chimes well with Polanyi's narrative too.

The environmental narrative must be understood as a way to organize facts, processes, and events as representations of a confined period of time into a configured sequence of completed actions (Cronon 1992). Many anthropologists, ethnologists and human geographers who build their research upon old paradigms such as the spatial turn in the sociology of the 1980s (Sheller 2017) and the more recent cultural turn in geography[2] still rely on such narratives because they contain a normative dimension that is not up for debate. As Cronon argued, environmental narratives should not just be about stories of nature but include stories about stories of nature. Otherwise, they turn into fictions converting dynamic and open processes

---

[2]The cultural turn in social geography with its claim to be a distinctive post-disciplinary approach based on critical political economy and the analysis of capitalist social formations (Barnes 2001) does very much follow the narrative of the dependence—and neo-Marxist theories of the 1970s with their unarticulated normative dimension (Aerni 2011c).

into petrified accounts with a beginning, a middle and an end. As such narrators of environmental stories are to give their story a teleological form, selecting some contextual facts that fit the story and ignoring others. As such, they are able to endow their story with a sort of moral authority that links past actions to the present and a predicted future, with their preferred intervention to restore the original state. The problem of such storytelling is not necessarily what is said, but what is excluded from ensuring the compelling moral argument of the story (Aerni 2011c). It turns everything that remains unmentioned into something that does not deserve the attention of moral authority because it is deemed inauthentic or illegitimate. In addition, since the 'local' (seen from a global perspective) is often associated with something authentic and moral, whereas the opposite attributes are assigned to 'global', the clear distinction between the global and the local in sustainability research often serves as a pseudo-scientific foundation for philosophical idealism and political voluntarism (Scott 2017).

One common denominator of *environmental sciences*, dealing with ecosystem models, *economics*, dealing with equilibrium states, and *human geography* (understood as the spatial analysis of the economic, social and environmental dimension of a geographically defined region) is that they all dispense with a contextual and dynamic understanding of sustainability that is open and non-judgmental. Quite often, the narrative in human geography is not made explicit but instead concealed behind a scholarly language that borrows from a mechanical and static understanding of ecosystem dynamics.

According to Rangan (2000), the story could go as follows: Being geographically isolated from other regions of India, the peasant societies of pre-colonial Garhwal were mostly homogenous, without hierarchies or class divisions. The causal factor for the current misery is not primarily the modernizing Indian state, but the somewhat colonial rule, which initially established the foundations for a formal economy designed to exploit the forests and other natural resources ruthlessly. In other words, the ecological and social decline in the Garhwal region is attributed to an external force that led to exploitation, population growth and eventually environmental degradation and poverty. So the implicit assumption is that the Garwhal region forms a clearly defined socio-cultural and environmental ecosystem that experiences detrimental disturbance from outside in the form of companies engaged in the commercial exploitation of its resources. The only way to restore the original state is then to push the national government of India (the proper external counterforce?) through international declarations and national protest campaigns to pass laws that ban external stakeholders from affecting local people and their environment. The return to the fictitious original state of blissful self-sufficiency did not take place in the Garhwal region. Rangan argues instead that the environmental and social situation deteriorated even further because the 'good' external stakeholders (civil society) used local communities in their discursive strategies as a means to an end to please their constituencies elsewhere. The content of the discourses themselves has mostly been de-coupled from the local stories and the concerns expressed by the local people.

4.3.3 'Epistemic Brokers' in Postmaterial Societies: The Case of Vandana Shiva

The successful discursive strategies that have the power to mobilize at a global scale often involve former grassroots activists who are able to speak the language of the locals and know how to portray the local concerns to a global audience in a way that is meaningful and well-digestible. These actors are also called 'epistemic[3] brokers' (Herring 2010; Kolins Givan et al. 2010; Freidberg 2007). They have successfully established knowledge claims on the global level that do no more account for the local context or the geographical history in which the knowledge has developed initially. As such, epistemic brokers create a global epistemic community that regards them as a credible source of knowledge in global networks ensuring the diffusion and perpetual repetition of the same knowledge claims until they reach the status of objective knowledge validated through a system of belief, acceptance, verification, and justification.

This allows epistemic brokers to accumulate a considerable amount of cultural capital from which they can eventually make a good living. As global celebrities who are identified with a morally compelling global environmental narrative, they have also become prominent 'influencers' in the social media (Himick and Brivot 2018). As such, they were able to transform local resistance movements in defense of concrete material interests into a new social movement largely concerned with postmaterial values such as the environment, local identity, and gender.

A seminal article in the journal 'International Organization' by Haas (1992) described the power of expert networks that shape the way policy decision makers frame problems, and formulate policies. As such, these expert networks would have assumed control over knowledge and information in political decision-making processes.

Epistemic brokers may once have challenged such expert networks. Yet, today, they have created powerful belief systems on their own that are actually replacing the previous expert networks in politics. Rather than directly influencing policy decisions through government advice they influence them indirectly by shaping public opinion and, with it, exert external pressure on policy decision makers. The rise of social media has accelerated the diffusion of their ideas and their claims to the extent that they create new patterns of beliefs and frames of interpretation on a global scale that remain mostly unchallenged (Nichols 2017).

In the case of the Chipko movement, Rangan attributes this achievement to epistemic brokers such as Sunderlal Bahaguna and Vandana Shiva. On the national level, Bahaguna was able to combine the Indian government's rhetoric on national self-sufficiency, development, and national cultural and economic self-defense with the symbolic acts of popular protest. On the global level, the epistemic broker of the

[3]Epistemic (derived from the ancient Greek word episteme or knowledge) refers to the process in which knowledge is created, validated, justified and disseminated in society to create systems of 'rational' beliefs.

Chipko movement was Shiva, the anti-globalization activist. She was able to mediate between local/national and global networks by providing global symbolic meaning to the local Chipko movement. Instead of portraying the women in the Garhwal region who protected their trees by embracing them as activists defending their economic right to make commercial use of their own forest resources, she argued that these women have no desire to engage in the cash-based economy. As such, she attributed to them a larger moral purpose related to the struggle against big corporations and in favour of cultural and ecological self-determination. As a consequence, the Chipko movement and, with it, the Western tree-hugging movement became globally celebrated as the vanguard for the preservation of ecosystem services as well as a symbol of non-violent protest and the feminine principle of nature (Rangan and Garb 1996).

What Shiva essentially achieved was the conversion of a local story into a global myth by portraying the Chipko movement as a manifestation of the struggle between two paradigms. The 'life-enhancing' paradigm based on the feminine principle and subsistence-oriented natural resource management, and the 'life-destroying' paradigm derived from the masculine principle and its drive to destroy the natural resource base for economic gain. To erase any form of ambiguity, Shiva prefers to speak on behalf of these local women rather than inviting them to speak on their own behalf. She refers to them always as 'they', but it is never clear who 'they' actually are. Sentences such as 'they expect nothing from development' or 'they want to preserve their common property resources and subsistence base against commercial interests' can be read in one form or another over and over again in her books. Ultimately, 'they' became the symbol of resistance against a disembedding global economy. Why bother if 'they' have no clue what Shiva is talking about?

In other words, Vandana Shiva managed to shape a global understanding of local concerns in India that bears no relationship to facts on the ground anymore. Her postmaterial narrative completely ignores that the indigenous people in the Himalaya never lived in complete isolation, but were always integrated into a supraregional economic and political system that provided them with useful goods and services. The core of this system were far-reaching markets where local people could sell their goods and find employment during off-season. Ultimately, the myth about the Chipko movement constructed on the national and the international level has influenced Indian politics more than the local story of resistance. Protecting the pristine forest ecosystem of the Garhwal region became the priority, not the people that lived in this ecosystem and made use of its resources.

Yet, rather than becoming self-critical about all the unintended consequences, Shiva discovered the business of being a global epistemic broker who is able to convert all kinds of local resistance movements driven by concrete material interests into a black-and-white identity movement, driven by postmaterial values and the struggle against negative external forces such as MNEs and modern technology in agriculture. As a result, she quickly diversified her protest portfolio while sticking to the overall theme of the local with its attributes 'small', 'traditional', 'female', 'self-sufficient', 'ecological', 'natural' versus the global with the attributes 'large', 'modern', 'male', 'profit-driven', 'exploitative', 'unnatural'.

In this context, the term 'GMOs' (genetically modified organisms), applied to food and agriculture, became a very suitable target of global protest because it could easily be attributed to all the negative external forces associated with 'Big Business' that threaten traditional and sustainable lifestyles (self-sufficient communities) in agriculture. There is hardly any urban food, seed, earth or food sovereignty festival in Europe or the United States where Shiva does not show up as a prophet who denounces in inflammatory terms the current 'life-destroying' practices in agriculture and promises hope regarding a future free of the commercial interests that would produce only 'slavery' and 'rape' in farming. As a prophet of postmaterial society, she gets well-paid for speeches and flies first class (Specter 2014). Moreover, her Navdanya farm in the north of New Delhi has become the spiritual Mecca for 'life-enhancing' agricultural practices, sponsored and visited primarily by Western environmentalists and aristocrats.

Her rhetorical trick to convince people to fight against the use of GMOs in developing countries is very similar to the one she applied to make people believe that the tree-hugging women in the Garhwal region were representing the feminine principle that enables communities to live in harmony with their natural environment (Rangan and Garb 1996). Again, she talks on behalf of the alleged 'victims' but instead of the tree-hugging women, the word 'they' applies to innocent small-scale farmers in India that have been duped allegedly by Monsanto to adopt genetically modified (GM) Bt[4] cotton seeds, which would have then caused environmental problems, indebtedness, misery and ultimately farmer suicides. Again, a myth with no need to ask the more than seven million farmers, occupying 10 million hectares, who have adopted the technology in India, representing nearly 90% of all Indian cotton fields (Specter 2014).

Have all these small-scale farmers been so gullible as to buy the 'seeds of death'? As innocent and vulnerable victims who need protection from outside agents to restore the original harmonious state, peasants in India are not meant to have any agency on their own. Their judgment does not count because they are either naïve or brainwashed by Monsanto. That is how the urban environmental myth, as shaped by Shiva, goes.

Lots of recent findings indicate however that small-scale farmers in India are not as gullible as Shiva portrays them after all. Empirical studies in the field found no evidence that GM cotton increased suicide rates (Gruère et al. 2008). Since the seed was introduced, yields have increased by more than 150% according to a study conducted by the International Food Policy Research Institute (IFPRI) (Gruère et al. 2012). Moreover, there are indications that GM cotton also benefited the health of farm families and improved the education of their offspring since less labor-intensive spraying was necessary, which used to be done by family members, including the children in the absence of means to hire landless workers

[4]Bt stands for Bacillus thuringiensis (Bt), a soil bacteria that synthesizes different insecticidal proteins named Cry, Vip, and Cyt that are able to kill different insect orders, or nematodes. These proteins have been extensively used in insect control practices in agriculture, including in organic agriculture.

(Krishna and Qaim 2012). No one denies that pests have developed resistance against Bt cotton in regions where the adopting farmers have not received any technical assistance. Still, the problem of resistance development is not new. It exists since the beginning of the history of plant breeding and is not specific to GMOs. It can even be observed in organic agriculture where Bt is sprayed since it is a natural soil bacterium. Moreover, the damage caused by pests developing resistance may not be as high as often portrayed in the Western media since Indian farmers have a choice of 800 different genetically modified Bt cotton hybrid seeds (Specter 2014). Moreover, the Ethiopian government approved the commercial release of Bt cotton in Ethiopia in 2018, not because Monsanto bribed them but because a high-level government delegation visited India in 2017 to talk to farmers who have adopted Bt cotton.

The spread of Bt cotton in India also has very little to do with the power of Monsanto. Shiva's argument that the patents of the US company Monsanto prevented poor farmers from saving seeds in India is wrong: the Indian Plant Variety and Farmers' Rights Act of 2001 denies patents on plants and guarantees every person the right to "save, use, sow, resow, exchange, share, or sell" seeds.

India has one of the weakest systems of intellectual property rights protection in plant breeding in the world, which explains why most semi-subsistence farms rely on saved seeds from the previous season to produce their food crops; not because they prefer them, but because there is no choice.

4.3.4 Cultural Appropriation and Denial of Local Agency

Epistemic Brokers such as Vandana Shiva are very familiar with terms that are fashionable in 'culturally sensitive' Western society. One of them is 'cultural appropriation', which has a moral undertone stemming from disapproval of the copying of elements of the minority culture by the dominant majority culture. This disapproval is particularly strong if the use of appropriated cultural elements is completely detached from the original cultural context and if members of the originating culture have not been consulted (Hartigan 2005; Scafidi 2005). For example, fashion designers and artists have faced criticism and even 'shit storms' in the social media for appropriating elements from cultures considered 'vulnerable'. Appropriators are accused of reducing cultural elements with spiritual and ceremonial significance in the original society to 'sexy' costumes and 'exotic' accessories.

Cultural appropriation deserves shaming when people merely care about the nice-looking artifacts of a particular ethnic group but take no interest in the lives of the people who represent them. Alternatively, when companies commercialise cultural artifacts without having received prior informed consent and without any benefit-sharing agreement with the originator (which is in most cases already illegal).

However, the view that an ethnic group would generally condemn the commercial use of its cultural artifacts may again be a claim of global activists that has not been validated. Most ethnic groups are eager to engage in cultural and economic exchange and also tend to integrate certain foreign elements into their own native culture. What matters is that there is the mutuality of respect and mutual benefit (Avins 2015). The idea that culture is something authentic and pure and therefore should not be diluted or copied reflects the bipolar world view that prevails in affluent urban NIMBY communities. They take pride in being culturally sensitive by pointing out that they purchase cultural artifacts from fair trade shops and visit cultural events in the city that celebrate multiculturalism and tolerance, presumably in ways that are free of cultural appropriation. After all, they merely support the preservation (the allegedly authentic performance) and not the alteration of the culture. Apart from trying to be culturally sensitive, members urban NIMBY communities are also concerned with the quality of life in their respective neighbourhoods. Yet, they prefer to share their quality time with like-minded people which may conflict with their openness for other cultures. They would like to 'help' people from other cultures, migrants for example, as long as they comply with the stereotype of the classic victim who shares the view of a hostile neoliberal world and gratefully accepts their help, though simultaneously respect their private sphere (meaning 'do not enter my backyard'). After all, there is a justified fear that the real people may not confirm the romantic stereotypes about their culture and therefore lead to disappointment (Lander 2008).

Romantic stereotypes may have once been created by colonial merchants, officials, and adventurers who returned to Europe to tell people about the half-naked exotic tribes they had discovered. Nowadays, epistemic brokers concerned with manufacturing stereotypes tend to originate from the urban cultural elite of the 'exotic' developing country, but have studied in a Western country and therefore are very familiar with the mindset of urban NIMBY communities in affluent societies. They claim to speak on behalf of the self-sufficient traditional communities in their respective countries, which are portrayed as innocent, in harmony with nature, egalitarian and highly vulnerable to 'cultural appropriation' due to their inability to defend their own cultures. The message of this portrayal is very Polanyian in the sense that it suggests traditional communities to live in a happy and self-sufficient state until outside commercial interests corrupt local values and norms and alienate the natives from their own culture.

Yet, this view implies that cultural identity has never been confronted before by cultural and economic exchange but instead has grown out of the native soil. It also assumes that the people involved in these traditional communities have only one wish, namely to be left alone; and, in case there is already ongoing cultural appropriation, they would fight for the recognition of cultural ownership but without any interest in making commercial use of their cultural artifacts. This is the language that resonates well with people in affluent societies because it suits the NIMBY narrative well—'please live your lives, we live our lives'. As illustrated in Sect. 4.3.2, the Chipko movement never defined itself as a stand-alone culture with no interest in commerce. Furthermore, their attitude towards nature was utility-oriented, rather than merely value-oriented or spiritual as portrayed by the epistemic brokers. In other words, cultural appropriation takes place not so much by Western commercial interests but by civil society activists in developing countries to earn a good living from a Western constituency that admires and supports them.

By siding with the external actors and their focus on merely protecting and preserving local culture, epistemic brokers tend to betray their own people and actively contribute to the decline of tradition, language, habits, norms and values. After all, a large part of the culture will not be passed on to the next generation because the offspring is likely to migrate in view of the growing external material dependence and the risk of becoming mere museum objects. In other words, they are unwilling to play a game in which they have no agency.

4.3.5 The Temptation in Academia to Uncritically Embrace Environmental Narratives

Epistemic brokers may have to fear the anger of their own people, but they can be quite confident in increasing their power and prestige on the global stage. However, their opportunistic behavior in politics and their indifference towards insights from empirical research and statistical facts does not deter academia from honoring them as advocates of global environmental and social justice.

Vandana Shiva again serves as a case in point. As a 'scholar' of sustainable agriculture, alternative livelihoods, ecofeminism, cultural self-defense, etc., she has received honorary doctorates from several universities in North America and Europe and won the Sydney Peace Prize for her commitment to social justice and "her tireless efforts on behalf of the poor". Beloit College, in Wisconsin, even honored Shiva with its Weissberg Chair in International Studies, calling her "a one-woman movement for peace, sustainability, and social justice" (Specter 2014).

How is it possible that academic institutions have become so uncritical in their praise of an environmental activist who helped create divisions where there should be collaboration? Her polemic campaign has discouraged farmers from combining the best tools of precision agriculture and biotechnology with the practices in agro-ecology and participatory plant breeding. The divisive view, propagated by Shiva and her allies, is not compatible with the UN Sustainable Development Goals that call for global partnerships instead of a war against business, science, and technology.

The current failure to come up with constructive solutions to global challenges in academia may have a lot to do with the fact that interdisciplinary and practice-oriented collaboration in the social sciences has become very difficult. As Rangan puts it, interdisciplinary social science research has become risky and difficult because "it requires drawing on vocabularies and concepts in ways that recognize the disciplinary conventions and traditions" (Rangan 2000: xiv). The interdisciplinary approach goes beyond the loose collaboration of like-minded experts in the different disciplines. It involves a mixed method approach that starts with inductive research, in the form of field investigation designed to gather information on the ground and listen to different stakeholder and expert viewpoints. Based on this first phase, hypotheses can be developed and subsequently tested in a deductive way by means of quantitative methods (Aerni 2011c). The argument that critical social science research must contain an inductive, as well as the deductive part, goes back to social science theorist John Dewey (1905). He insisted that interpretation of data requires translation and effective communication between disciplines as well as between people and places. It also consists of the art of letting local people speak for themselves while refraining from interpreting their views uncritically in the light of a global narrative.

According to Rangan (2000), interdisciplinary and critical social science research also requires moral courage because it challenges the discursive power of the incumbent epistemic communities in academia. To pursue an academic career outside of these epistemic communities has become all but impossible. She puts her faith in independent research best described as political ecology, with consideration of context and geographical history. Political ecology should refrain from superimposing political ideology in the interpretation of the findings (as it often happens in human geography). It should also dispense with the popular mythological narratives of Polanyi and Shiva and instead enable the communication of a more complex reality of local development.

The approach is very much compatible with the integrative approach pursued in the field of economic sociology. Economic sociology has, however, the additional advantage of providing a more compelling theory addressing the contextual and historical dimension of sustainable change based on the concept of embeddedness, referring to the fact that there are no economic relations without prior social relations. The following chapter illustrates how such integration of different social science disciplines could take place around the basic insights of the field of economic sociology.

Chapter 5
The New Understanding of the Term 'Embeddedness' in Economic Sociology

In the 1980s, sociologists started to pursue a more empirical approach to embeddedness by focusing on the social networks that underlie economic interactions. As such, they returned to the questions that mattered to classical sociologists like Emile Durkheim, Georg Simmel, and Max Weber, namely how the evolution of social institutions takes place in capitalist economies. The term 'embeddedness' was not applied anymore in Polanyian terms to distinguish between fictitious substantivist (embedded) and formalist (disembedded) economies, but was used instead to define the concrete, ongoing systems of social relations that underpin economic action (Granovetter 1985).

Mark Granovetter, the sociologist who played a crucial role in re-defining the term 'embeddedness', started with a general criticism of over- and under-socialized conceptions of human action in sociology and economics. While the over-socialized conception in sociology mainly saw societal norms and values, internalized through socialization, as explanatory factors of human interaction, neoclassical economics started from an under-socialized conception with its homo oeconomicus who has no age, is not embedded in any social structure and completely independent of social relations in economic decision-making. According to Granovetter, actors do not behave or make decisions outside a social context. Nor do they act exclusively according to prescribed roles that reflect a sort of moral norms and values. Instead, it is social relations rather than merely institutional arrangements, as argued by Acemoglu and Robinson (2013), that directly account for a low degree of fraud and malfeasance and the production of trust in economic life. His approach to embeddedness concretely follows and analyses patterns of social relations to explain how societies overcome the problem of trust and disorder in economic life (Granovetter and Swedberg 2001).

P. Aerni, *Global Business in Local Culture*, SpringerBriefs in Economics,
https://doi.org/10.1007/978-3-030-03798-7_5

5.1 The Moral Dimension of Entrepreneurship

Granovetter is in line with the argument of the economist Albert O. Hirschman, who pointed out that the pursuit of economic self-interest in the formal market economy has hardly ever been based on reckless utility-maximizing attitudes designed merely to realize material short-term benefits. Instead, human behavior in formal markets is characterized by civilised human interaction based on meta preferences related to the respective cultural upbringing and social relations (Hirschman 1992). Also, each economic actor faces the risk of reputation loss, if he fails to play by the unwritten rules of kind interaction. In this context, Hirschman refers to the so-called "doux-commerce" hypothesis, as suggested by well-known philosophers of the age of enlightenment such as Montesquieu and Condorcet. The hypothesis asserts that people involved in supra-regional formal markets are more likely to develop civilized manners in dealing with each other and have a more profound sense of fairness and reciprocity than people who remain in economic autarky. This hypothesis is confirmed in recent empirical and experimental research in the field of behavioral economics (Henrich 2006).

5.2 Embeddedness as a Way to Address Three Major Coordination Problems

By focusing on micro- rather than macro- foundations of the concept of embeddedness, Granovetter leaves Polanyi's normative macro-narrative unchallenged. In this context, Beckert (2007) proposes to revisit Polanyi's work starting from the basic non-normative insight that, ultimately, it is the embeddedness of economic exchange that makes economic and social integration possible. The challenge, then, is to identify the socioeconomic preconditions that allow economic prosperity to be combined with the realization of a humane social and political order.

The reduction of uncertainty is an indispensable precondition for the emergence and operation of market economies. In this context, the embeddedness of economic exchange in a web of social and legal relations helps to reduce this uncertainty and provides stability to market interactions. Beckert (2007) contends that the embeddedness of economic interaction manifests itself in three types of coordination problems that need to be addressed to reduce uncertainty to a level that makes economic exchange possible: the problem of value, the problem of competition and the problem of cooperation.

He believes that the importance of economic embeddedness can only be understood in the context of these three coordination problems, especially when economic exchange moves from local to global business interactions. Beckert discusses them as a fundamental concept without going into much detail on how coordination is to be addressed by MNEs that intend to do business in high-risk developing countries with significant cultural differences and great institutional uncertainty. The following three sub-chapters illustrate how foreign investors usually negotiate value, competition, and cooperation in their particular industry with local stakeholders. These negotiations may result in arrangements that enhance the benefits of FDI for local communities and the environment while enabling the MNE to build up the necessary social capital to reduce the problem of asymmetric information and, with it, the transaction costs of doing business in a foreign country with substantial institutional, cultural and economic differences. It illustrates well how global business needs to embed itself in local culture to secure its long-term license to operate. It requires an understanding of context and geographical history as well as the willingness to listen to and understand the concerns of local stakeholders.

5.3 Value as a Coordination Problem

The problem of value refers to the fact that the price of a product, reflecting its current demand and supply in the market, may not correspond to its social value taking into account the potential of the product to address societal needs.

The valuation of a particular product is primarily social in character and, as such, generates preferences that are endogenous (from within society). The value problem is most apparent when comparing the uncontroversial price-setting strategies of companies that operate in the market for consumer electronics or luxury goods, and the highly controversial one in the field of life-saving drugs or productivity-enhancing technologies in agriculture. Whereas no one bothers if the company Apple asks for a higher price of its latest smartphone, society may disapprove of companies that generate profits by selling life-saving drugs that should be accessible to everyone. If such products are not accessible to those in desperate need because they cannot afford to pay its price, pricing becomes a moral issue for society. At the same time, such products would not exist in the first place, if companies had not developed them in the hope of creating a profit that is at least able to reimburse the fix costs invested in R&D and the cost of going through the regulatory approval process (Kock and Gould 2013). So whereas the price of a smartphone is mainly based on quality markers influenced not just by advertising but institutions and platforms that discuss its merit and thus the value of the product, the social process of valuation of products that are generally associated with public goods, such as life-saving medicine (no one should be excluded, its consumption should be non-rival), show that the embeddedness of economic action makes it necessary to negotiate prices in a way that make them socially acceptable as well.

Often, corporations may do so on their own through a strategy of price differentiation (Ridley 2005). The pharma industry may generate substantial profits in affluent countries where generous health care systems rather than patients represent the main buyer of their products. This provides companies with the necessary incentives to sell the same products at a much lower price in low-income countries, as long as there are efforts to prevent parallel imports. Yet, price differentiation strategies are increasingly branded by retailers and activists in affluent countries as an unfair rip-off (why pay double the price of a particular life-saving product in Switzerland compared to neighbouring Germany?)

The conflict between companies that would like to set the price of their new life-saving product at a level that ensures a decent return on investment, and poor people who need access to it, is often addressed through governmental intervention. Governments act as bulk buyers of the particular product for their country. As such, they can negotiate favourable terms and thus ensure improved access for its citizens. However, the question that remains unanswered is why all people in a developing country should benefit from the much lower price, independent of their economic status. After all, these countries also have many dollar millionaires who could easily pay the price that they would have to pay in an affluent economy. An individualized pricing system based on the idea of fair compensation combined with an open-licensing approach would be the most effective solution from an equity perspective to address this challenge. However, politically, it may be difficult to find a majority for such an approach because the public debate on IPR is highly polarized with business on the one side insisting on the validity and usefulness of the existing IP system, and civil society on the other side calling on free access to everything by everyone in the name of the right to food, right to health and right to well-being (Kock 2018).

Yet, this polarized global debate is often de-coupled from genuine local concerns, as illustrated in the previous chapter. Local stakeholders who deal with global stakeholders willing to invest in the local economy may be less interested in ideology than in concrete results and therefore more willing to embark on new solutions that benefit both parties while encouraging further open innovation.

5.4 Competition as a Coordination Problem

The problem of competition relates to the fact that 'perfect competition' may be neither in the interest of business nor the interest of society as a whole. Companies do not like to enter markets characterized by perfect competition because the profit margins are tiny and returns are likely to decrease in the course of commodification. In other words, perfect competition is where the marginal revenues meet the marginal costs, which means that there is no profit left to be reinvested in business (Romer 1994).

Society as a whole disapproves of perfect competition as well because it is associated with a cut-throat competition that only allows those players to survive in the market of price takers that manage to produce cheaper by saving on labor costs. It has been argued that perfect competition and its exclusive focus on efficiency gains may, therefore, lead to a decline in product/service quality, worker safety, and environmental protection because costs tend to be saved first in areas where they do not hurt short-term profits (Shaikh 2016).

Since perfect competition does not generate the necessary corporate profits to invest in innovation, its long-term value for business and society is minimal. While innovation may create large positive social externalities, investing in innovation requires a company to have assets generated from previous rent- or profit-seeking activities. Moreover, a company would only invest in innovation if there is a prospect of generating a future profit through a temporary monopoly, in which no other firm is selling the new product, which gives the company temporary price-setting power (Romer 1994).

This explains why markets of perfect competition are either fictitious or in rare cases an ephemeral phenomenon that is eventually undermined by three forces: lobbying by market incumbents for regulation that prevents market entry of new competitors; labor, environmental and consumer pressure groups that advocate regulations that make it more challenging to save costs at the expense of society and the environment; and, finally, future-oriented companies that want to grow through innovation. The latter asks governments to create an institutional setting that rewards them for taking the risk of investing in something new. Patent protection is one but by no means the only way to enable innovative companies to have a temporary monopoly that allows them to generate profits that cover for the prior R&D costs and enable further investment in the continuous improvement of the product (Matthews and Zech 2017). In other words, the demand to regulate markets of perfect competition is meant to discourage cut-throat competition and encourage more investment in innovation instead. After all, it is innovation that generates the profits that also make it possible to offer favourable employment conditions and to improve the environmental sustainability of business activities.

However, the creation of barriers to market entry that shield incumbents from perfect competition can also harm long-term societal interests, if the monopoly position has not been achieved on temporary terms through innovation, but instead through lobbying activities aimed at shutting out competitors in a particular market. Companies that achieve such a monopoly can extract a rent from customers, not by offering a better and more innovative product, but by simply raising the price of their existing products given the absence of more competitive market players. This leads to a welfare loss for consumers without any compensating welfare gain resulting from innovation. Such state-supported monopolies become insiders that benefit from the status quo and have little interest in change and economic renewal (Olson 1982). That is why they often embrace a NIMBY rhetoric portraying themselves as defenders of the local economy and local culture against profit-seeking foreign predators. This stance endears the companies to the public and strengthens their position in the political bargaining process (Aerni 2015b).

Governments should become more alert regarding these disguised attempts at regulatory capture by the local incumbents of an industry. Instead, they ought to play the role of facilitators that encourage long-term foreign investments in local companies that focus on innovation rather than lobbying. The purpose of their facilitator role is to provide incentives to entrepreneurs to create new markets by helping them to reduce the risk of investing in new products and services that may generate increasing returns with considerable external benefits for society (Romer 1994; Aerni 2015b).

5.5 Cooperation as a Coordination Problem

There is no competition that does not build on prior cooperation designed to set the informal or formal terms under which competing actors pursue their self-interest. Constitutions of democracies, as well as market institutions, are products of such prior cooperation. They aim to ensure that the pursuit of self-interest in business and politics is rule-based and does ultimately benefit and not harm society at large (Buchanan and Tullock 1962).

The coordination problem of cooperation arises from asymmetric information and the uncertainty of expectations and intentions among exchange partners in business and society. Informal local business activities cope with this problem by relying on social networks in general and trusted ethnic and kinship networks in particular. In tight-knit networks, the cost of cheating and defection are high because they lead to the exclusion from the supporting social relations (Hirschman 1970). In global formal business relations, it is more challenging to enforce informal rules of fairness and reciprocity because the costs of defection are lower unless the economic actors involved decide to engage in pre-competitive cooperation to agree on the terms of doing business. Moreover, a third party institution has to be appointed to enforce the terms. This third party must be endowed with sufficient enforcement power that is able make a credible threat of exclusion from business to the extent that defection becomes costly (Olson 1965; Hindmoor 1998). There is therefore not just public but also a private interest in regulating business.

For example, as discussed in Sect. 4.2.4, prior to the spread of public food safety standards in the 20th century, there were private food safety standards designed and implemented by the food industry through an arrangement of pre-competitive collaboration (Ping 2011; Freidberg 2009). The reason why all parties in an industry were prepared to be inspected by a third party was the fact that deceiving the consumer, for example with diluted milk or false weight measurements, did not just affect the business of the cheater but led to a loss of trust in the whole industry. The resulting systems of reputation helped to overcome the problem of asymmetric information. When private standards eventually became national and international public standards, they could be effectively enforced only in countries where the coordination problem of cooperation in the private sector has previously been addressed adequately by private standards (Ping 2011; Aerni 2013b).

Institutional uncertainty in developing countries is mostly related to the fact that these countries have not been part of the first wave of globalization in Europe and the United States when industrial sectors learned to define and enforce rules and norms on their own. Instead, they directly adopted public regulation as recommended by their formal colonial powers after World War II. Since this public regulation could not draw on prior experience of a self-regulating local industry, the external imposition of formal rules tended to be merely ignored by the local private sector. Local people in business resented the rules imposed by foreign powers and preferred to stick to their own informal set of rules with their specific enforcement mechanism rooted in strong cultural embeddedness (Henrich 2011). In other words, the formal rules imposed from outside lacked legitimacy because local stakeholders had not been involved in defining these rules.

As a consequence, MNEs that aim to do business with actors in the local economy must first try to understand and respect the local rules through cultural embeddedness. At the same time, they must also comply with global industry standards in their particular value chain and abide by formal codes of conduct. Merely adjusting to the local terms of business may lead to undesirable embeddedness associated with collusion and corruption, which puts subsidiaries at odds with their corporate value system and lead to reputation loss abroad.

Addressing the problem of cooperation, therefore, requires a corporate commitment to principled embeddedness, understood as an active engagement with the local stakeholders to learn about local needs and expectations, and to explain the formal institutional constraints of the Global Value Chain (GVC) in which the MNE operates. Local actors that would like to become part of these integrated GVCs as suppliers to the MNE may have to learn the global language of business and its formal operations. In this context, MNEs must be willing to assist their potential local partners in complying with the necessary formal requirements (timely delivery of pre-defined quantities, compliance with social and environmental standards, product quality and sanctions in case of malpractice) through training and capacity development programs. In view of the fact that this is time- and cost-intensive and does not belong to the core business activities of the company, professional firms or local industry associations have assumed coordinating function by assisting local companies in their efforts to upgrade their business to an extent that they meet the conditions to become suppliers of MNEs (Watkins et al. 2015). The integration into formal GVCs benefits the local company in the sense that it acquires the ability to attract investment and sell its products beyond the neighbourhood area. It will also help the local company to expand its market share in the formal domestic economy that is requiring compliance with public regulation. Through repeated exchange with the MNE, the local company learns to trust the foreign stakeholder and respect its business principles. This, in return, enables the MNE to earn social capital and, with it, local legitimacy. This reduces uncertainty and lowers transaction costs.

Beckert (2007) considers social legitimacy a precondition for the emergence of stable markets. The definition of what is legal and what is illegal to buy and sell is determined by society and its norms and values rather than the economic actors (Jacobs 1970). However, norms and values change over time and with it the definition of legal and illegal transactions (Appiah 2011). Even though illegal markets may work like legal markets, the way the coordination problem of cooperation is addressed is different. Loyalty is of existential importance to the illegal networks; if one party defects under the promise of impunity, the whole illegal business network may be dissolved and its members end up in prison. This also explains the strong embeddedness of economic activities in kinship-based criminal networks such as the Mafia (Gambetta 1996). Therefore, if the underlying norms and values of the respective network are not beneficial, but harmful for society as a whole, they must be discouraged, and MNEs should refrain from getting involved by all means (De Sardan 1999).

Even though extensive research has been conducted in the field of experimental economics and psychology on trust and cooperation (Gintis and Bowles 2005; Henrich and Henrich 2007), the field does not take into account how economic and technological change can alter the terms of cooperation. Beckert (2007) points to the standards of creditworthiness that evolve as social and economic conditions change and new technological possibilities arise. New applications for smartphones and biosensors, for example, allow inspectors to detect non-compliance of local suppliers concerning social and environmental standards more cheaply and effectively, while the inspectors are being tracked to ensure that they actually visit the sites. The same technologies help local suppliers to meet the standards at much lower costs. As a result, effective cooperation in developing countries may more easily move from interpersonal relationships in tightly knit cultural networks of social exchange to more anonymous relationships with stakeholders that may not be based on the same values and norms that shaped the original local network of trust, but share common interests instead (e.g. avoiding accidents and hazards). However, the global networks of cooperation will not and should never replace the local networks because it is the local networks that ultimately underpin the social legitimacy of the global ones. The mutual dependence but also the benefits resulting from enhanced exchange between global and local networks eventually leads to hybrid and purpose-oriented network structures that do not necessarily undermine but actually complement more traditional networks.

5.6 Embeddedness in the Context of Economic Complexity

Recent research in economic complexity (Hidalgo 2015) mostly confirms the importance of economic and cultural embeddedness of business activities by pointing at the purpose-oriented social networks created at a considerable cost to address context-specific coordination problems in long-term global collaborative networks of firms. They provide the foundation for the adoption or even creation of innovation in global value chains (GVCs). The links in these purpose-oriented networks are highly specific and recurrent, and the concrete outcome is not known in advance. They involve individuals and firms that must be understood as packages of specialized knowledge and knowhow nested in a social structure that supports them. Such global networks designed to create a stream of revenue out of economic interaction are challenging to manage and maintain because they are evolving and, with it the knowledge and know-how, they embody. Accumulating knowledge and know-how is difficult because creating the networks required to embody them is very challenging. If they succeed in achieving their purpose and become economically prosperous, it is less because of standardization and technologies that lower the costs of creating formal network links, but because they are embedded in pre-existing cultural and social networks that are based on social capital, the foundation of trust required for starting an initial informal collaboration. Since losing trust has serious long-term consequences for the one who committed a breach of trust, this non-contractual, informal form of collaboration is handy in deterring fraud and malfeasance (Shapiro 1987; Molina-Morales 2010). At the same time, it works without the burden of lengthy legal documentation and costly enforcement procedures, and, as such, makes it worthwhile to embark on risky joint commercial projects. This also explains why most employers are keen on hiring referrals; after all, it is easier to trust someone with whom you share a friend than someone who is not connected to anyone you know (Granovetter 1995). For MNEs that intend to invest in a developing country where they have no established network yet, referrals of qualified people familiar with the local culture and well-embedded in social networks, often constitute the first entry level. Since such cultural brokers are familiar with the global and the local, they may then lead to the establishment of social relations of the MNE's local subsidiary and eventually provide the foundation of formal business relations.

Understanding how complex and purpose-oriented global networks work, requires knowledge of transaction cost theory, which deals mostly with formal network links. Besides, it must be based on an understanding of the concept of embeddedness derived from economic sociology, which focuses on the investigation of informal links (Hidalgo 2015). The economic effects of pre-existing social networks are essential for understanding the costs of establishing the initial informal links.

Chapter 6
Economic Globalization as a "Disembedding" Force?

As the main drivers of economic globalization, MNEs are crucial in addressing coordination problems in GVCs by tapping the knowledge and know-how of their embedded internal and external networks to generate economic outcomes that ensure a continuous stream of revenue through scalable innovation, but also benefit the societies in the regions in which the MNE invests (Andersson et al. 2002). Yet, in public, they have not been associated with embedded economic structures but rather with the disruption and displacement of locally embedded traditional economies (Robinson 2014; Baker 2016; Shiva 2016).

MNEs may indeed be a major force of economic globalization that has induced many locally embedded traditional industries to either transform or disappear. But, as discussed before, this pressure on local economies started already in the 19th century in Europe, and abuse was always part of it. For example, the landlords of the rural feudalist system in Eastern Europe wanted to compete in the new trade-oriented market economy, not by introducing new forms of division of labor and technological innovation, but by merely squeezing out more work obligations from their dependents. Fernand Braudel called the new arrangement a 'second serfdom' (Braudel 1982) that led to an economic as well as a humanitarian crisis. Indentured peasants finally had an opportunity to exit the exploitative system by escaping overseas. The average migrant arriving at the end of the 19th century in the United States may not have belonged to the poorest of the poorest, because they simply lacked the means to move overseas. Yet, the fact that only 10% of the arrivals from Poland were literate and skilled is an indication that the majority may have escaped serfdom (Zeitz 2017). The 'New World' enabled them to take their destiny into their own hands.

Eventually, feudal entrepreneurs benefiting from the old exploitative manorial system proved unable to compete with the rising bourgeois entrepreneurs who became more productive and innovative, not by exploiting their employees but by investing in them (Holenstein et al. 2018).

The plan to exit the old manorial system was a risky and uncertain venture for the former servants that migrated elsewhere. But it eventually allowed them to

P. Aerni, *Global Business in Local Culture*, SpringerBriefs in Economics,
https://doi.org/10.1007/978-3-030-03798-7_6

unshackle the old fetters associated with traditional feudal economic repression and to flourish in arrival cities in Europe and overseas. Therefore, none of the former subjects may have deplored the disembedding and eventual disappearance of the old manorial system during the first wave of globalization. It may undoubtedly have caused new uncertainties and new ways of legal and illegal exploitation, but for the first time, it also created unprecedented economic opportunities for outsiders in society, those who had no economic rights under the previous rulers (Aerni 2016).

6.1 Why Disembedding Traditional Structures May Help Outsiders

The great migration of the underprivileged in the 19th century to new places with better economic opportunities jumpstarted a process of social mobility also in Europe. It forced the previously unchallenged aristocrats to deal on equal terms with entrepreneurial upstarts who emerged from the former peasant and servant class (Braun 1965). Often the latter did not compete directly with the former who invested primarily in established capital-intensive industries. Instead they created entirely new markets addressing old needs in a new way (Etzkowitz 2004; Aerni 2007a). The new markets had to resolve Beckert's three fundamental coordination problems of embedded economies (see Sect. 5.2): determining the value of the new product, securing a profit through an acceptable departure from perfect competition, cooperating with other stakeholders on issues of common interest.

Thanks to their ability to limit fraud and malfeasance in practice by adequately addressing the coordination problems of cooperation, these new economic actors created the necessary social capital to gain a public license to operate in new and often untested fields of business (Cope et al. 2007; Aerni 2016). Their success in creating and scaling new markets allowed many of these upstarts to eventually lay the foundation for the creation of a multinational corporation (MNC).

However, as the landed gentry before them, many emerging large corporations may have used their enormous profits, originally generated through innovation, to merely imitate the lifestyle of the aristocrats through conspicuous consumption. Instead of investing in the future, they secured their rents by optimizing the management of their assets and by spending on political lobbying to ensure high barriers to market entry (Schluep and Aerni 2016). Moreover, even in the early stage, when they were innovators and therefore created considerable external social benefits, they may have also caused substantial external social costs, in the form of worker accidents and consumer health hazards, for which they were unwilling to be held liable or assume responsibility for the damage (Ice 1991).

Emerging large corporations may also have posed a risk to embedded traditional societies without offering any positive alternative to the originally very repressive system. In this context, Polanyi (1944) rightly refers to the growing anger of the public against reckless corporate behavior and how it eventually led to counter-reforms that improved social security and health coverage for the employed and helped create social safety nets for those who lost their employment or were unable to work in the first place.

But this view mistakenly implies that disembeddedness always happens through 'bad' private sector investment while re-embeddedness takes place exclusively through 'good' government intervention designed to repair the damage caused by the private sector. This implicit view is also very much present in Dani Rodrik's understanding of embedded liberalism (Rodrik 2011). It assumes that opening up markets to international trade has to be accompanied by the expansion of the interventionist state. In other words, the state will obtain the approval for more open markets in a democratic system only if it protects and compensates potential losers. Even though it is indeed true that the private sector may not be able to provide for public goods, the public sector is highly dependent on private sector knowledge and technology in the adequate provision of critical public goods. Moreover, the public sector needs to raise sufficient taxes from successful private sector activities in the first place in order to finance a modern welfare state (Desai 2003). Also, one has to take into account government failure in the sense that governments may provide access to public resources only to those loyal to them, as it happens in traditional patron-client systems. Such governments fail to provide public goods even if the funds are available. Finally, many of the knowledge-based private sector innovations eventually end in the public realm and thus assume the character of a classical public good. They become freely accessible for anyone interested in commercializing generic versions of the original innovation (Naam 2013). Therefore, the provision of public goods by governments should generally not be understood as a correction for the damage done by private sector activities, but as a result of the emergence of a formal, tax-paying private sector that is most productive and innovative when operating in an open market system that is free of collusion, discrimination and corruption. A prosperous and innovative private sector also contributes to the effective management of public goods through its competences in the provision of client-oriented products and services (Aerni 2015b).

6.2 Disembedding Post-Colonial Structures

The social and economic transformation through economic globalization in Europe during the 19th century is to some extent comparable to the transformation of Asia in the 21st century. FDI by MNEs in many Asian countries has challenged and sometimes disembedded the prevailing traditional rent-seeking economies inherited from colonial and pre-colonial regimes. These former regimes had many similarities with the feudalist economies of the Ancien Régime in Europe. Colonial powers were rarely interested in disrupting existing economic and political systems in the countries they colonized. Instead, they preferred to collaborate with local elites based on the principles of reciprocity, redistribution, and autarky that prevailed in the traditional embedded economy (Aerni 2007a). Many Asian countries were able to overcome this collusion of the colonial and the domestic elite, which went at the expense of the local people.

Alas, this collusion still prevails in many African countries. Appiah (2006) points at the roots of this collusion by illustrating how the recruitment of slaves in Western Africa to serve the plantations in the new colonies worked. Slave trade was mainly based on the pre-market principles of gift exchange between local tribe leaders and European slave traders. Nelson Mandela also illustrates in his autobiography 'Long Walk to Freedom' (1994) how traditional tribal leaders in South Africa resisted his efforts to end the Apartheid regime because the regime run by European settlers supported and helped to secure their tribal power in the self-governing Bantustans. European settlers promised not to challenge their authority as traditional leaders over their territory, as long as they ensured that their subjects do not try to migrate to their cities in South Africa.

Finally, the Cold War period in the 20th century further strengthened the feudalist economies in developing countries as either client states of the United States or the Soviet Union (McCormick 1995; Zubok 2009). Both powers endowed the new indigenous leaders of the European ex-colonies in Africa with almost dictatorial powers while securing their indirect rule with military bases and propaganda about the merits of their economic system (Dunning 2004).

With the end of the Cold War and the fall of the Berlin Wall, such systems of patronage lost their purpose and faced growing criticism from within and without the developing world. Moreover, the economic rise of Asia illustrated that developing countries are not entrapped in an exploitative global economic system in which they merely serve as suppliers of raw materials. In this context, Asia's success falsified all the neo-Marxist and structuralist theories of the 1970s with their implicit simplistic assumption that 'we are rich because they are poor'. Alas, due to path-dependence in the social science disciplines and the fact that falsification in the social sciences does not work in the way it works in the natural sciences, the paradigm shift did not take place as a result of new insights gained from recent history (Aerni 2011c).

6.3 How FDI Contributed to Catch-up Growth and Economic Empowerment

Asian Countries like Japan, South Korea, and China may have resented political interference of Western powers, but they very much realized the value of foreign direct investment (FDI) from the West. After all, if Western multinational corporations find it attractive to invest in their countries under the condition that they become partially embedded into their traditional economies (e.g., through joint ventures and the creation of special economic zones), they are likely to contribute to the transfer of important know-how and technology. This transfer enabled the domestic economy to reach a level of sophistication that allowed it to successfully compete with Western economies in the global market (Yingming 2009).

It illustrates how long-term FDI leads to unprecedented catch-up growth (Aerni 2011b, Romer 2010) and the emergence of an entrepreneurial middle class, which makes use of its economic empowerment to lobby for political reforms that are more conducive to their economic interests (Woetzel 1989).

Catch-up growth rates today are much higher than they used to be in earlier times since so many more goods and services are around and so much more know-how, technology, and financial means are in circulation today. These tangible and non-tangible goods are only waiting to move into places that previously have been partially or entirely isolated (Romer 2010). Even minor reforms of the institutional setting in low income countries with high transaction costs can quickly transform a traditional rent-seeking economy into a more prosperous market-based economy that allows for social mobility and the emergence of an entrepreneurial middle class. Some scholars, therefore, argue that the real root of inequality is the fact that certain governments decide to deprive their people of the benefits of globalization while others encourage global economic integration to benefit from catch-up growth (Hidalgo 2015). Thus global inequality is not merely the result of the accumulation of capital by a global finance elite, but maybe even more related to flawed government policies.

In this context, government policies that either completely cut off the domestic economy from international trade or completely open up to trade, can be equally harmful for low-income countries. Merely lowering barriers to international trade in existing goods and services with decreasing returns to capture efficiency gains from trade will not make a domestic economy per se more prosperous even if comparative-static general equilibrium models used by trade economists would suggest so. After all, economic prosperity is the result of the creation of new markets with increasing returns rather than mere efficiency gains resulting from the liberalization of commodity markets with decreasing returns. New markets with increasing returns emerge by making use of the non-rival resource knowledge to improve or even create entirely new goods and services that better meet particular needs in society. In this context, knowledge must be understood as a form of immaterial good represented by non-tangible assets such instructions, designs, protocols, and recipes that can be shared and copied without losing in value. In fact, unlike material goods that lose value with increasing use, knowledge gains in value the more it is used. Depending whether knowledge is used to improve products in the domestic economy or whether it is applied in export-oriented industries, it may or may not result in more foreign exchange through trade (Romer 1994). However, a domestic economy that is highly diversified thanks to the continued adoption of new knowledge and know-how in different industries, is also more likely to increase and diversify its exports (Hidalgo 2015).

Most of the basic and applied knowledge, used by entrepreneurs to produce new goods and services, is found in the public realm. It has contributed to a global stock of knowledge that can be tapped to embark on catch-up growth and make a more sustainable use of scarce natural resources provided that countries pursue policies that encourage the adoption and commercial use of knowledge. Deaton (2004) attributes the gap between encouraging gains in life expectancy and the somewhat disappointing record in GDP growth per capita in many developing economies over the past fifty years to the fact that the spread of useful knowledge (formula to make a medicine) may have contributed more to improved living conditions than the mere import of physical goods (e.g. pills). The example often used is Oral Rehydration Therapy (ORT). ORT helps to replace the water and electrolytes lost through diarrhea. The formula is simple but very effective in curing life-threating forms of diarrhea. WHO's diarrheal disease control programme, established in 1978, played an important role in popularizing this treatment throughout the world. Overall, it is estimated that the ORT treatment saved the lives of more than 50 million children during the 25 years after it was popularized in 1978 (Fontaine et al. 2007), not by importing pills but by enabling local doctors to make use of the instructions to create the mix of water and electrolytes to treat the illness effectively on site. If one tracks the diffusion of ideas, in this case related to the spread of the instructions to prepare an ORT treatment, it becomes clear why the increase in life expectancy turned out to be the bigger success story than the increase in income. The ability to harness the global flow of ideas to make local goods and services in other domestic economic sectors may therefore not just help to close the gap between life expectancy and income but also contribute to inclusive growth, one of the primary objectives of the Sustainable Development Goals (SDGs) of the United Nations.

6.4 'Knowledge', an Underused Resource in Efforts to Cope with Environmental Challenges

Economic policies designed in the 21st century should not rely on the definition of economics as the science of the optimal allocation of scarce resources because knowledge is a resource that is not subject to the laws of scarcity (Warsh 2006). New knowledge applied in the production of goods and services is therefore crucial for coping with the global sustainability challenges related to the decreasing availability and quality of scarce and rival environmental goods. After all, it is new knowledge that leads to scalable innovation, which makes it possible to produce more and better output with fewer inputs. It, therefore, should not be treated anymore as a 'residual variable' in Total Factor Productivity (TFP), but should instead be recognized as the crucial input factor in a dynamic global knowledge-based economy (Jones and Romer 2010). Yet, policymakers tend to listen to established economists who were trained during the Cold War period, often out of convenience or reverence. Their theoretical mindset and global outlook tend to be very different from the practice-oriented view that prevails in the domestic entrepreneurial middle class. As a result, policies relying on the old recipes of neoclassical economics tend to be rather unpractical. This is particularly true in low income countries, where an entrepreneurial middle class is still largely missing and therefore not able to oppose policies that do not encourage home-grown economic development. Apart from relying too much on foreign advice, their governments address the preferences of powerful landlords and traditional leaders because they are more likely to ensure their re-election than entrepreneurs (Aerni 2006). As patrons, these landlords and traditional leaders ask their dependents or clients to vote for the ruling party in return for protection of their material interests and improved access to scarce government resources (Berman 1998). This may not please donors that care about 'good governance'. However, government officials have become very skilled in wrapping their interest in preserving traditional patron-client systems into a language of caring and solidarity with the poor. This language resonates with foreign aid experts who deplore the vulnerability of embedded local economies. Moreover, what donors with professional staff ultimately care about are national progress reports on poverty reduction, food security, climate change, gender, etc. Such reports are tangible and can be presented to interested parties as proof that something is being done for aid money. Governments in low income countries, therefore, employ their most skilled people for the creation of such reports.

6.5 The Failure of Foreign Aid to Empower Local Entrepreneurs Through Economic Integration

The long-term impact of development assistance (DA) has not been positive in most African countries (Deaton 2015). The sobering record of foreign aid may be related to the entrenched principal-agent problem in official DA, embodied in the fact that aid organisations are accountable to donors in OECD countries rather than the poor in recipient countries (Aerni 2006). In other words, foreign development agencies are primarily focused on addressing the preferences of the taxpayers and donors of the country that sponsor their activities—rather than the preferences of the poor in the recipient countries.

The preferences of donors and taxpayers in economically matured economies tend to be similar to the preferences of the aristocrats of Europe in the 19th century. Affluent people increasingly resent economic globalization because it challenges their economic and political supremacy. Moreover, why should they recognize the need for change, if it is unlikely to make them better off? As a consequence, they prefer projects that protect traditional communities from the forces of economic globalization rather than support the endeavour of such communities to become better integrated into a supra-regional economy (Aerni 2016).

This also applies to government agencies in OECD countries in charge of implementing 'aid-for-trade' projects. The aid-for-trade initiative was launched at the Sixth WTO Ministerial Conference held in Hongkong in 2005 and was meant to promote supply-side capacity and trade-related infrastructure, particularly in low income countries. Developing countries feel however that the initiative has not delivered what it promised because donor countries decided to disburse 'aid-for-trade'-funding through existing channels of DA. Ultimately, it may have created more and better-paid jobs in donor countries than in recipient countries (Hallaert 2017).

6.5.1 Self-Serving Nature of Swiss Sustainable Trade Promotion and Development Research

In Switzerland, the Swiss State Secretariat for Economic Affairs (SECO) largely used 'aid for trade' funding to promote 'sustainable trade,' which assigns sustainability labels to a wide variety of tropical agricultural products such as coffee, cacao, and banana shipped to Switzerland (Robin and Wyss Bisang 2017). The Swiss sustainable trade initiative mainly benefits Swiss and Dutch NGOs that are meant to train farmers in tropical countries to comply with Business-to-Business (B2B) and Business to Consumer (B2C) standards. Especially in the case of B2B standards that have previously been designed and implemented by the respective industries involved in the production and trade of agricultural commodities, it is questionable whether delegating the capacity development to NGOs and public sector agencies adds value to farmers.

A survey with producer organizations and stakeholders involved in extension services conducted by the Tropical Agriculture Platform (TAP) based at the Food and Agriculture Organisation (FAO) of the United Nations in Rome showed that, in terms of extension services, private sector capacity development for agricultural innovation is preferred to public sector/civil society initiatives by the local stakeholders (Aerni et al. 2015). This is relatively easy to understand since farmers who are trained by agribusiness companies that directly buy their products have a more clear idea about the skills and technologies that growers need to acquire in order to comply with the required sustainability and quality standards.

One of the more significant players in Europe which claims to make sustainable trade mainstream in agriculture, is the Dutch organization 'IDH'. On its website[1] it presents itself as a convenor of companies, civil society organisations, governments and others to create effective public-private partnerships. Farmers and their know-how and experience do not seem to matter much however. Dutch (BUZA), Danish (DANIDA) and Swiss (SECO) government agencies fund IDH. It can safely be argued that such organisations together with the numerous mostly European-based certification agencies and NGOs that partner with them, benefit the most from such government-funded sustainable trade initiatives (Aerni 2017c).

Over the past decade, SECO supported several B2B initiatives in agricultural commodities trade. But one-third of the funding for sustainable trade is still used to support cooperatives in tropical countries that produce certified fair trade and organic products, so-called business to consumer (B2C) labels for Swiss retailers.

[1]See https://www.idhsustainabletrade.com/about-idh/ (Accessed Sept 25 2018).

This is a sort of indirect subsidy for these cooperatives in the tropics as well as the Swiss retailers that aggressively advertise B2C labels as the 'moral' choice, not least, because they still earn the best margins on these premium labels. It is therefore not surprising that neither retailers nor the cooperatives in the tropics that benefit from the more stable and slightly higher commodity price as well as Swiss government assistance, have much interest in upgrading standards and making organic and fair trade more open to innovation and entrepreneurship. Its effect of preserving existing structures rather than enabling structural change is best known in Swiss agricultural policy with its system of direct payments. The system was meant to promote multifunctional agriculture but proved to be somewhat ineffective in addressing social and environmental concerns in rural areas because it tends to undermine rather than encourage innovative entrepreneurship. Farmers are primarily concerned with ensuring the privileged access to scarce public resources (Aerni 2009).

Fair Trade and Organic labels seem not very concerned about their tendency to cement unsustainable structures in developing countries rather than enable sustainable economic change in the local economy. Why fixing something that works well for all parties involved? The problem is that these islands of allegedly morally superior agricultural production are subsidised and controlled by stakeholders abroad and tend to be wholly detached from the surrounding local economies. They, therefore, contribute little towards inclusive growth, economic empowerment, and sustainable innovation, which are some of the main pillars of the sustainable development goals (SDGs), as will be discussed in more detail in Sect. 7.4.6.

Another problem with buyer-driven B2C standards is that they may not promote inclusive agricultural trade but actually become an obstacle to it. After all, these standards have the effect of non-tariff trade barriers that cannot be disciplined by the WTO since member states, not retailers signed its agreements. The WTO Trade Facilitation Agreement (TFA), approved in Bali in 2013, should have had a significant impact on the priorities set by national agencies in charge of disbursing aid-for-trade funding. Yet, the established stakeholders involved who benefit from 'aid-for-trade' ensured that the priorities remain the same (Hallaert 2017). Trade facilitation programmes are primarily add-ons to existing projects. In the case of Switzerland, the contributions to international, mostly UN-based, trade facilitation initiatives are mostly described by SECO in terms of inputs. But whether the output proves indeed valuable to solve bottlenecks of technical trade barriers in developing countries and increases trade remains uncertain.

The defensive understanding of agricultural sustainability in affluent societies is self-serving. After all, it is them who currently benefit from the preservation of the status quo and have reason to fear that they may lose their privileged position with the current rise of the emerging economies. This self-serving aspect is often concealed behind a language of moral concern about all those people in developing countries who might become disembedded in the course of economic globalization (Aerni 2011b).

As a consequence, the aid priorities of official DA are focused on preserving cultural and economic practices in the informal economy in these countries, portraying their survival strategies back home as lifestyles that are sustainable, because they are focused on self-sufficiency rather than growth and the pursuit of gain, as Polanyi would describe it. These priorities in DA have also influenced the funding priorities of academic research in the area of poverty reduction, environmental management, job creation and food security in developing countries. The Swiss Programme for Research on Global Issues for Development (r4d programme), a joint funding initiative by the Swiss Agency for Development and Cooperation (SDC) and the Swiss National Science Foundation (SNSF) illustrates this well. The programme was endowed with a total budget of CHF 97.6 million for the duration of 10 years (from 2012 to 2022).[2] The projects funded are all based on a defensive understanding of sustainable development. They range from various types of agro-ecological and organic agriculture projects to classical aid for vulnerable communities (access to health, technology for disabled people, gender issues) to promoting climate resilience and monitoring and the promotion of peace and good governance. Not a single project was funded that deals with the role of the private sector in enabling inclusive growth.

As with the previously described Swiss sustainable trade initiative, the r4d programme claims that the selected research projects would reflect the priorities of developing countries. Yet, fact is that Swiss institutions must submit the research project proposals.

For Swiss applicants to the r4d programme, it is quite easy to find partners in developing countries who agree that the suggested initiative is of relevance and reflects a local priority because getting funding from a Swiss institution transforms the lives and the material conditions of the beneficiaries immediately and fundamentally. It often happens that researchers in developing countries know at a very early stage about the priorities outlined in r4d calls for proposals as well as the crucial topics that are likely to get funding. They may then initiate the application process by contacting Swiss researchers to submit a joint proposal that is likely to get funding.

The r4d programme further claims that the goal is to promote innovation. Yet, not a single project deals with the promotion of local growth-oriented entrepreneurs that ultimately account for commercially viable innovation. One wonders therefore how this SNSF programme contributes to the SDGs which puts emphasis on inclusive growth and innovation.

---

[2] See overview of funded projects: http://www.snf.ch/de/fokusForschung/nationale-forschungsprogramme/nfp-73/seiten/default.aspx.

6.5.2 How the Anti-business Rhetoric of Epistemic Brokers Supports Incumbents

The narrative of the poor but happy people that are in danger of being exploited by MNEs is popular in DA and very much shaped by the rhetoric of 'epistemic brokers' who claim to speak in the Global North on behalf of the poor and exploited in the Global South. Yet, as illustrated in Sect. 4.3.3., epistemic brokers do not reflect the priorities of the reformist forces in domestic politics of their home country, but rather represent the interests and ideologies of identity-based movements on the far left as well as on the far right. They often share the common goal to prevent economic change because it may dissolve hierarchical structures that secure their privileged access to public resources and economic rents.

Reformists often challenge the interests of traditional leaders because they resent the traditional patron-client systems that prevail in embedded informal economies and hamper economic development. Instead, they prefer to effectively address the three major coordination challenges of economic embeddedness in order to improve the attractiveness of the country for FDI. Their goal is to manage the transition from informal to formal economic activities through economic reforms that enable structural change, as it has been envisioned by the noble-prize winning Caribbean-born economist Arthur Lewis (1955).

Ignoring the demands of reformist forces makes foreign donors often part of the problem since they tend to neglect the needs of the educated and entrepreneurial outsiders in the domestic economy. Facing a dearth of political and economic support, these outsiders cease to envision a future in their rent-seeking economy and seek instead a better life in domestic cities. If the economic opportunities of these cities are also hampered by burdensome regulations that serve incumbents, the only option is to migrate abroad or to exit the system altogether. This has happened in the case of those young Arabs who committed suicide by self-immolation in despair of the absence of any economic rights that would have allowed them to make a living. With their tragic action, they jumpstarted what is generally known as the 'Arab spring'. Its primary purpose was to get rid of the old rent-seeking autocracies, rather than striving for an Islamic state. Even though all these events are well documented (Martin 2012; De Soto 2011), they have been largely ignored by the international donor community.

6.5.3 Local Growth-Oriented Entrepreneurs as Drivers of Economic Integration

There may be some timid initiatives to promote entrepreneurial rights and entrepreneurship, but these initiatives are focused mainly on improving livelihoods in the informal sector.[3] Such efforts have little impact on the overall economic situation, or employment opportunities because operating in the informal economy does not allow an entrepreneur to grow and formally employ people, as Arthur Lewis has observed (Lewis 1955). Why? Because only formally registered businesses that comply with standards and regulation and are able to reach a customer base beyond their neighbourhood, attract investment to develop a growth-oriented business that generates increasing returns and creates jobs in the formal sector (Aerni 2015b).

Enabling informal businesses to succeed as growth-oriented formally registered companies requires competent assistance and mentoring, which is found in the private sector rather than government or civil society (Aerni et al. 2015). However, unlike donor agencies and NGOs, companies invest in local businesses only if they are confident that the business will eventually succeed. In other words, an investment must pay off; and in case it pays off, it will not just benefit the respective local business and its investor but the region as a whole through its contribution to endogenous (home-grown) economic development with all its positive regional spillovers.

In this context, subsidiaries of MNEs that are committed to local embeddedness have the potential to contribute significantly to sustainable change in the region in which they invest. This requires, however, that governments and civil society groups do not erase business opportunities in regions they have identified as 'vulnerable' by creating a parallel economy dedicated to providing goods and services deemed to improve 'local resilience'. This would crowd out private sector investments and consequently increase dependence on external support while undermining the process of economic empowerment (Farla et al. 2016; Ciaian et al. 2012).

A recent paper by Amendolagine et al. (2017) was able to provide empirical evidence of the benefits of embedded FDI in developing countries in terms of sustainable economic change and inclusive growth. The authors found out that more intense GVC participation and upstream specialisation are associated with a higher share of inputs sourced locally by foreign investors. These effects have had a crucial impact on poverty reduction, economic empowerment through local entrepreneurship, and the creation of high-quality jobs. Yet, countries with weak law enforcement capacity and low-quality education systems benefited less from the positive impact.

[3]See for example the UNDP Initiative on livelihood and jobs: http://www.undp.org/content/undp/en/home/sustainable-development/development-planning-and-inclusive-sustainable-growth/livelihoods-and-jobs.html (Accessed Sept 25 2018).

Chapter 7
Embedded MNEs and Their Contribution to Sustainable Change

Since the UN published the Guiding Principles on Business and Human Rights (UNGP), numerous MNEs have embraced the Principles on a voluntary basis, and several UN member states have enacted national action plans on business and human rights as part of the State's responsibility to disseminate and implement the UNGP.[1] Some of these action plans even contain legally binding reporting requirements for MNEs. The problem with the UNGP document is that the guidelines assume by default that MNEs and their subsidiaries may try to violate human rights and destroy the environment unless they fear to be held accountable for their misdeeds. This baseline assumption runs against the basic insights in new economic sociology and economic history discussed in Sect. 5.1.

Throughout history, companies that recklessly pursued their economic self-interest at the expense of public well being hardly ever thrived in the long run because selfishness is not a good way to build up the social capital necessary for the creation of lasting business relations. Most companies have in fact an active interest in addressing the coordination problems of cooperation, competition and value (see Sect. 5.2) with local stakeholders. This approach to cooperation includes the creation and enforcement of commonly agreed industry standards as well as codes of conduct designed to discourage fraud and malfeasance. Moreover, most companies tackle the coordination problem of competition by lobbying for relatively detailed social and environmental standards that ensure high barriers to market entry for cheaper competitors with lower societal and environmental standards. In addition, addressing the coordination problem of value includes negotiations on the 'fair price' for an innovative product or process in the area of food and medicine with governments from developing countries to contribute to improved access, as well as the design of open licensing platforms that provide researchers with access to many technology

[1]See http://www.ohchr.org/EN/Issues/Business/Pages/NationalActionPlans.aspx (Accessed Sept 25 2018).

P. Aerni, *Global Business in Local Culture*, SpringerBriefs in Economics,
https://doi.org/10.1007/978-3-030-03798-7_7

research tools—free of charge for non-profit institutes. These advances in institutional arrangement and the mutual interest in sharing technologies have contributed directly or indirectly to social, humanitarian and environmental improvements in developing countries (Scherer and Watal 2002; Kock 2018).

7.1 Coping with Business Coordination Problems Through a Strategy of 'Principled Embeddedness'

A corporate commitment to 'principled embeddedness' is primarily designed to address the coordination problems of the market with local stakeholders jointly and responsibly. In this context, MNEs based in high- and middle-income countries investing in low-income countries realize that they have to abide by global industry principles and guidelines, even if domestic regulation and weak legal enforcement power in host countries may offer ample opportunities to cheat. MNEs with a long-term investment strategy tend to be committed to principled embeddedness, not because of a normative ethical stance, but because it is in their enlightened self-interest to be perceived as a fair player in business and society, independent of the location in which they operate. At the same time, MNEs know that they do not operate in a vacuum. The local stakeholders in the region in which they invest will eventually raise the issue of how much the presence of the MNE benefits the local economy as well as local society and the environment. In this context, there may be a divergence of what the MNE thinks is of importance to the local community and what really matters to the local community. A local materiality assessment may therefore be of crucial importance in setting the right priorities in the design of a local stakeholder engagement strategy.

There may also MNEs that are determined to merely extract local resources to serve people elsewhere. As such, they may turn into economic off-shore islands with hardly any links to the local economy and no social impact in the region whatsoever. This could, however, backfire because nationalist politicians may eventually denounce the MNE for not caring about the host country and its people. As a result, the foreign investor may lose its license to operate or be forced to comply with somewhat rigid local content requirements or joint-ventures that may make little sense from a business point of view and consequently make foreign companies more reluctant to invest.

So, no matter if they pay their taxes in the country in which they operate and comply with international business and human rights standards, MNEs will have a local acceptance problem, if perceived as generating profits for stakeholders abroad while shunning any interaction that might benefit the local economy and society (Aerni 2017b; Schluep 2017). Therefore, MNEs with long-term investment strategies have an interest in negotiating with local stakeholders on the conditions required to integrate qualified employees and local business partners into their global supply chains. For that purpose, they may have to invest in local capacity development and training programs necessary to recruit more skilled local employees and source from local companies. Since they may be unwilling to carry the burden all on their own, they must also encourage the government to play a supportive role by tailoring the corresponding public educational and vocational training programs more to the needs of employers. Governments that create a favourable institutional environment will enable subsidiaries of MNEs to lower their transaction costs. It will also make it easier for MNEs to engage with local culture and build up social capital. Such a joint commitment to 'principled embeddedness' therefore essentially contributes to the SDGs and should be encouraged correspondingly by the UN member states that endorsed them including Agenda 2030, designed to map out the implementation of the Goals by 2030.

7.2 The UNGP and Its Potential Conflict with Principled Embeddedness

The focus of the UN Guiding Principles on Business and Human Rights (UNGP) on 'do no harm' may help discourage corporate fraud and malfeasance of MNEs in vulnerable economies. But it may also discourage a corporate commitment to local embeddedness in these economies. Business partners in the local economy are not used to, and maybe also not in a position to comply with costly international social standards. Moreover they often run local informal side businesses that are difficult to monitor. This may increase the reputation risk of an MNE even if the local business partner is compliant with standard due diligence requirements in the bilateral business relation.

Moreover, bringing prosperity to the people and helping the region to become more sustainable is unlikely to be attributed to an MNE's commitment to principled embeddedness as long as the public debate, and the academic discourse in law, economics, and political science, is framed in a way that attributes private benefits strictly to the private sector and public benefits strictly to the public sector (Aerni 2015b). This strict division of benefits stands in contrast to the fact that the long-term improvement of living conditions is first of all due to a better average household income as a result of private sector development (Juma 2011). Yet, governments and international organizations hardly mention this basic insight because it would undermine the claim that any improvements in living standards are a result of their public policy initiatives.

The focus on risk avoidance advocated by the UNPG and the low likelihood that better local living conditions will be attributed to FDI tend to increase the costs of MNEs to invest in 'high-risk' developing economies.[2] As such they also discourage

[2]The term 'high risk' is derived from the OECD country risk classification; see http://www.oecd.org/trade/xcred/crc.htm (Accessed Sept 25 2018).

corporations to commit themselves to principled embeddedness. Whatever business the MNE is pursuing in such a country, it is not able to avoid interaction with the respective government, which often represents the only political and economic powerhouse in the country. Being associated with the national government increases however the risk of being accused of complicity when it comes to all the unaddressed needs of the local population, government corruption and the lack of respect for democracy and human rights (Aerni 2017b). As a consequence, many MNEs prefer not to invest, unless there is a high prospect for massive profits that covers for possible damages resulting from reputation loss abroad or expropriation by the national government on the domestic front, as it is often the case in the mining industry.

OECD countries that approve of legally binding national action plans on business and human rights may therefore inadvertently discourage their MNEs from investing in 'high risk' developing economies, which, again, brings them in conflict with their commitments to comply with Agenda 2030 and the implementation of the UN Sustainable Development Goals (SDGs). After all, the SDGs put more emphasis on private sector involvement than previous environment and development accords (Scheyvens et al. 2016). They recognize that responsible and innovative companies have the potential to be agents of sustainable economic change and inclusive growth (Kolk et al. 2017)—not by merely avoiding risks but taking the risk of investing in a new market (Aerni 2017a).

7.3 The Role of Subsidiaries of MNEs in Developing Countries

So far, MNEs have been portrayed in this book as homogenous entities, all located in wealthy OECD countries and investing in poorer non-OECD countries. This is misleading insofar as many of the globally expanding MNEs originate in emerging, middle-income economies rather than high-income economies. MNEs from emerging economies tend to be encouraged by their governments to invest in low income, and increasingly also in high income countries (Luo and Zhang 2016). Instead of being monitored and sometimes shamed for investing abroad, MNEs in emerging economies, and China in particular, obtain active support from their governments in bidding processes for public procurement especially in low income countries, for example through pre-financing agreements (Brautigam and Gallagher 2014).

MNEs in OECD as well as emerging economies are not homogenous entities because their local subsidiaries abroad often enjoy a substantial amount of autonomy, and they frequently are as much involved with local stakeholders (external embeddedness) as with their parent companies (internal embeddedness) (Scott-Kennel and Michailowa 2016).

This double or multiple embeddedness (Meyer et al. 2011) exposes the subsidiary to internal and external institutional forces: MNEs which adapt their strategies and organizational practices to the local context are subject to material constraints imposed by the available local resources as well as institutional constraints resulting from the need to abide by the rules of the parent company in a local institutional environment that asks for more flexible arrangements. If subsidiaries manage to combine external and internal resources successfully, they become centers of excellence on their own and, as such, may benefit the host country through their knowledge base as well as their commitment to sustainability and collaboration with domestic business (Andersson and Forsgren 2000).

The new emphasis in business research on the crucial and increasingly autonomous role of subsidiaries of MNEs for local endogenous development has led to a growing skepticism toward the former "center-periphery" view, in which the firm-specific advantages of MNEs are developed and controlled by the parent company, while their foreign subsidiaries merely exploit these advantages in the local markets. This may still be the case at an early stage when an MNE starts to explore opportunities to sell its products in other countries. However, once an MNE selects a country to become an important production base, the importance and autonomy of the respective subsidiary increases, and, with it, the pressure to enhance embeddedness in the country (Andersson and Forsgren 2000). Another critical dimension of enhanced local embeddedness may be related to the potential of the digital age to change the distribution of competitive advantages across different categories of firms in global value chains (GVCs). Information and Communication Technologies (ICTs) have facilitated the emergence of new manufacturing technologies and cost-saving logistics that render the integration of small and medium-sized enterprises (SMEs) in developing countries economically more attractive because of the rapid decrease in transaction costs. Enhanced use of ICTs in GVCs also leads to advances in product modularity. They can act as a substitute for relational governance, even when cultural distance among partners is high. In other words, fragmentation and modularity may decrease potential cultural clashes with local partner firms (Alcácer et al. 2016). At the same time, they increase pressure to pursue a strategy of principled embeddedness in the host country because integrating local companies into global value chains requires them to adjust their business practices to international standards.

7.4 Selected Cases of 'Principled Embeddedness' of Subsidiaries of MNEs

In 2015, seven case studies on the embeddedness of Swiss-based multinational firms in developing countries have been carried out by researchers based at the Center for Corporate Responsibility and Sustainability (CCRS) at the University of

Zurich. The funding for the field research was granted by CTI,[3] the Swiss Federal Commission on Technology and Innovation, renamed 'Innosuisse' in January 2018. The overall objective of the research project was to measure the local embeddedness of multinational enterprises in developing countries[4] and their potential contribution to local sustainable development and inclusive growth. Based on the insights gained from the CTI project case studies involving MNEs such as Chiquita, Nestlé and Syngenta, and a project workshop on embeddedness at the University of Zurich,[5] in which the case of Bata shoes was discussed, the following chapter illustrates what embeddedness means concretely. The CTI research project was the first of its kind. The field research was conducted in collaboration with the selected MNEs as well as with local partner universities and external, independent stakeholders who were familiar with the respective MNE subsidiary. This allowed to compare the extent and quality of embeddedness as described by the company with the perception of its embeddedness by the local stakeholders. In addition to the advancement of this contextual understanding of embeddedness, the project team also designed a prototype of an online platform that allows MNEs to assess their embeddedness in the local economy and eventually compare it with the industry benchmark.[6]

Even though some shortcomings and challenges in the business of the selected country subsidiaries have been identified, also thanks to the external assessments, an implicit commitment to 'principled embeddedness' was identified in all case studies. Independent of the failures and scandals that all these large MNEs may be confronted with in the course of their operations in developing countries, it is worth pointing out what makes their commitment to principled embeddedness valuable for the local communities and the environment in the following subchapters.[7]

7.4.1 Nestlé Philippines: An Locally Embedded Company

Nestlé Philippines is the local subsidiary of Nestlé, a Swiss-based company. It was set up in 1911 in the Philippines, together with Anglo-Swiss Condensed Milk Company. After a brief suspension of operations during World War II, the company made a comeback under the name Filipro Inc. Moreover, Nestlé S.A. and San Miguel Corporation entered into a partnership to combine imports with local production

[3]CTI is the Swiss Federal Commission on Technology and Innovation (https://www.kti.admin.ch/kti/en/home.html) sponsors research at universities that are considered to be of commercial value in the private sector.

[4]See http://www.ccrs.uzh.ch/en/research/sustainable-impact/projekte/local-embeddedness.html.

[5]See http://www.ccrs.uzh.ch/en/News/embeddedness-panel-en.html.

[6]See http://www.keiccrs.ch/ (Accessed Sept 25 2018).

[7]The following case studies are discussed in more detail in a special issue of the Africa Technology Development (ATDF) Journal published in 2017. The issue is accessible online on http://atdforum.org/1967-2/# (Accessed Sept 25 2018).

resulting in the formation of Nutritional Products, Inc. (Nutripro). In 1962, Nutripro's first factory started operations in Alabang, Muntinlupa to manufacture NESCAFÉ. In 1977, Filipro, Inc. and Nutripro Inc. merged under the name Filipro, Inc. In 1986, Filipro, Inc. changed its name to Nestlé Philippines, Inc. In late 1998, Nestlé Philippines became a wholly owned subsidiary of Nestlé S.A., following the latter's acquisition of San Miguel Corporation.[8]

Today, Nescafé Philippines is not just sourcing its coffee beans from the Philippines but also processes and sells most of its end products domestically. Moreover, most of its employees in senior positions are Filipinos. As such the company is indeed embedded in the local economy and perceived by most Filipinos as a Filipino company. Simultaneously, Nestlé adheres to its CSR standards and has developed a NESCAFÉ Plan designed to ensure that coffee farmers grow viable, healthy crops, and that coffee farming remains sustainable. Together with the Rainforest Alliance (RA), the Sustainable Agriculture Network (SAN), and the Common Code for the Coffee Community (4C) the company makes sure it complies with internationally recognised sustainability standards.

As such, the case of Nestlé Philippines represents a great case of principal embeddedness. The company has not just created tens of thousands of jobs since its first subsidiary was set up hundred years ago, but also created valuable knowledge transfer into the local private sector. This helped to nurture a national economic ecosystem surrounding the production, commercialization and consumption of coffee. Its focus on complying with the strict social and environmental standards of Nestlé Global in recent decades helped to ensure that the commitment to embeddedness also increased the sustainability of its the local suppliers through the company's extensive capacity development program.

7.4.2 Syngenta's Contribution to Capacity Development for Agricultural Innovation

The Swiss-based agribusiness firm Syngenta (recently acquired by the Chinese firm ChemChina) illustrates the positive impact of a self-interested commitment to embeddedness. Civil society activists have often criticized the company for being an agent of industrial agriculture that would generate profits at the expense of clueless small-scale farmers in developing countries. However, the fact is that an innovation-driven agricultural company like Syngenta is unable to sell any products to farmers unless they prove to be beneficial in terms of farm revenues and improved natural resource management. In this context, the company has an active interest in investing in those small-scale farmers in developing countries that are determined to move from subsistence-oriented to business-oriented agriculture. For that purpose, it has built up a global stewardship network in collaboration with local

[8]See https://www.nestle.com.ph/aboutus/history (Accessed Sept 25 2018).

NGOs that recognize the potential of public-private partnerships for sustainable development.[9] The network aims at ensuring the safe application of means of plant protection and teaches an integrated farm management approach. Each year, Syngenta and its partners train around three million producers worldwide and help them to upgrade their agricultural productivity sustainably.

Growth-oriented small-scale farmers benefit from Syngenta's stewardship program by learning how to comply with national and international labor and environmental standards. This instruction allows farmers to produce food for the growing urban population worldwide while generating the necessary revenues to invest in the future of their offspring.

It is undoubtedly true that the upgrading of small-scale agriculture goes along with an intensification of agriculture that also benefits Syngenta's sales of input. However, this may not necessarily be a problem, as long as farmers benefit from training regarding the safe and sustainable use of such inputs, and field officers can effectively ensure compliance with sustainability standards through a mobile phone app called 'farm force'.[10] Yet, when Syngenta input products are also informally available in local agro-dealer shops, there is a need to find ways to trace them and ensure that they do not cause environmental and farmer health problems due to incorrect application. A feasible approach to address this challenge is to improve the monitoring of farmers that obtain their products in agro-dealer shops. This can be included in the sale of microinsurance, as exemplified by the Kilimo Salama scheme.[11] Kilimo Salama was developed initially by Syngenta Foundation and eventually achieved a successful exit by being sold to ACRE,[12] the Agriculture and Climate Risk Enterprise. ACRE is the most extensive agricultural insurance program in Africa, and the first agricultural insurance program worldwide to reach smallholders using mobile technologies. The company allows farmers to obtain crop failure insurance for an additional down payment of 5–10% of their total purchase in the Agrodealer shop. Thanks to a completely automated index-based insurance system, farmers are informed via mobile phone about the likelihood of their farm experiencing crop failure through drought or flood. If the likelihood has reached a particular threshold, farmers obtain their compensation payment also via mobile phone. The system is now being upgraded to allow for the monitoring of biotic stresses such as pest and disease infection and to offer a discount of the insurance premium in case farmers adopt sustainable practices, as certified by local extension officers.

---

[9]See https://www.vsointernational.org/fighting-poverty/where-we-fight-poverty/bangladesh/growing-together-in-bangladesh (Accessed Sept 25 2018).

[10]Farmforce helps smallholders gain access to formal markets by using mobile technology to make traceability and compliance an integral part of smallholder production and to redefine the relationship between growers, manufacturers, and markets; see http://farmforce.com/ (Accessed Sept 25 2018).

[11]https://www.syngentafoundation.org/agricultural-insurance-east-africa (Accessed Sept 25 2018).

[12]In April 2018, ACRE won an award for Best African Startup in Agtech at the AfricaDays in Paris.

Syngenta's principled embeddedness approach illustrates how the private sector can enhance the capacity of farmers to innovate and become more productive.[13] This is not surprising given the fact that companies are under much greater pressure to deliver value for farmers. They need to understand their context and come up with tailor-made solutions, whereas public sector officials may offer services for free but they are often outdated, less responsive to the actual needs of farmers and therefore less valuable for them (Aerni et al. 2015).

7.4.3 Bata Shoes: Creating Welfare by Taking Rather Than Avoiding Risk

The international company Bata illustrates how many benefits MNE subsidiaries may create for local people if they do not primarily focus on merely exporting their products to high-income countries but instead, tailor them to the needs of the domestic market. Bata started its business in the Czech Republic, where it became a prominent shoe manufacturer for the army during World War I. After the war, it survived the economic slump by focusing on the mass production of affordable shoes for ordinary people. It was able to produce cheaply because the employees agreed to a wage cut in return for other benefits such as substantial discounts on food, clothing, housing, and other necessities as well as an offer to become an associate of the company through a profit-sharing initiative. With the onset of communism in Eastern Europe, Bata lost most of its shoe factories in the Czech Republic due to nationalization. As a result, it moved its headquarters to Toronto, Canada and later Lausanne, Switzerland, with global expansion plans.

Where ever Bata moved, it was welcomed as an agent of change with a mission to sell affordable shoes of good quality to the less affluent. Often the company represented a challenge to less innovative incumbent shoe producers in their regions of investment. Incumbents resented Bata's ability to scale shoe production and, with it, expand the supply of shoes leading to lower prices.

During the Cold War period, Bata expanded its business to Africa, where it identified significant economic opportunities due to a largely unmet need for shoes. The company has established production sites in several African countries where it produces shoes for the local and regional market. The purpose was to scale up production so that shoes are affordable to the locals. Once again, the company faced the threat of nationalization when many African countries became independent and turned to socialism. In order to avoid the same fate it met in Eastern Europe, it decided to venture into politics to avoid expropriation. The threat of nationalisation did however not induce the company to abandon Africa as a production site. It was rather the opposite. The company decided to embed its activities even more into the

[13]See http://www.fao.org/in-action/tropical-agriculture-platform/resources/reports/en/ (Accessed Sept 25 2018).

local economies. As such, it ensured local employment and economic empowerment by sourcing most of the shoe components from local companies. Zimbabwe is the most recent example, where Bata decided to stay in the country despite the political and economic uncertainty associated with the nationalization efforts of the local ruler, Robert Mugabe, and the international boycotts that followed. One reason why the factory was able to continue operating in the country was its strategy of local sourcing, which led not just to a vast amount of social capital in the local population but also helped the company to save a considerable amount of foreign exchange (New Zimbabwe 2015).

People recognized the value of Bata not just as a local employer but also a local stakeholder concerned with the health, the education, the skills and the economic situation of the local population. This concern made it risky for Mugabe to treat Bata merely as a foreign intruder that seeks to generate profits at the expense of people in Africa. After all, most Africans grew up wearing Bata shoes on their way to school. They were affordable, of durable quality and comfortable. In other words, the sale of shoes did not just enable the company to reimburse its investment costs but generated great external social benefits for underprivileged people. There are still some political agitators, such as Yves Engler in Canada, who denounce Bata shoes for having fought against nationalization of the company in several African countries during the Cold War period,[14] but such campaigns do not resonate among Africans. They regard domestically produced Bata shoes as a local brand that can compete with cheap Chinese imports. The Chinese plastic shoes may be cheaper, but they are less suited to the tropical climate and long-distance walking. Even impoverished African families are prepared to buy a pair of Bata shoes for US$15 to protect their children's feet on their way to school. They are an investment over a lifetime. Leather shoes can be shared by siblings and passed on to the next generation, if properly taken care of. It is therefore not surprising that 'bata' is the word 'shoe' in many African languages.

Bata shoe production and use also have significant health and educational dimension. The feet of millions of people in the tropics are attacked every year by hookworms, due to walking around barefoot in an unsanitary environment. Wearing shoes can massively reduce the occurrence of such skin and foot disease. Besides, Bata recognized the value of ensuring that employees work in a healthy and safe environment, which subsequently also influenced the workers' private lifestyles, attributing more importance to hygiene. All the endeavours of the company to improve their employee's lives triggered the aspiration of local families for a better future. Many of the children of the workers have received a better education and have been driven by an entrepreneurial spirit to improve things in their region. Bata runs a website that illustrates such concrete cases of social mobility.[15] As for education, the company has not just set up vocational training

[14]See https://www.pambazuka.org/democracy-governance/bata%E2%80%99s-footprint-africa-dark-story-canadian-shoe-giant (Accessed Sept 25 2018).

[15]See http://batalife.com/our-stories/ (Accessed Sept 25 2018).

programs, but also elementary schools because it regards local human capital as the foundation of its highly decentralized global business activities. Content, literate and highly skilled local workers with career options within and outside the company help secure the long-term operations in the respective country and provide a sense of ownership to the local people.

7.4.4 The Responsible Entrepreneur and the Selfless Communist

Many of Bata's achievements are attributed to the vision and mission of Thomas J. Bata, who advanced the family business after World War II. He wanted the company to win public trust by serving customers and employees (Bata-Blyth 2008). In this sense, Bata may have shared the moral virtues of another great European entrepreneur of the 20th century who became an icon of responsible capitalism: Adriano Olivetti. Olivetti was convinced that a global company must have the social goal to become firmly embedded in the local socio-economic and cultural context (Lu et al. 2017).

Responsible entrepreneurs such as Bata and Olivetti realized that they are not just economic but also moral agents of change. They knew that their business could only survive and thrive in the long run if its employees, its customers and the society in which its business operates do not feel cheated, exploited or deprived of essential economic rights. In other words, the virtue of a successful entrepreneur may consist of his or her 'selflessness', in the sense that an entrepreneur has to permanently think of serving others rather than him or herself, in order to be successful on the long run. This insight may also explain, why the Chinese government continues to believe that its current capitalist system is compatible with the prior goal of communism to eradicate poverty and treat people equally. Dr. Norman Bethune, a Canadian medical doctor and convinced communist who effectively brought modern medicine to rural China and tirelessly treated sick villagers and wounded soldiers, was praised by Chairman Mao for his selfless commitment to the people in his essay entitled 'In Memory of Norman Bethune'. The essay had to be memorized by generations of Chinese people as the virtues that should guide a good communist. But they could also be the virtues of a good entrepreneur in the capitalist economy. When Bethune argues that people should not ask 'How much have you got?' but 'How best can we serve you?' (Allan and Gordon 2009), this call would very much resonate with the responsible entrepreneur who has no time for conspicuous consumptions but reinvests profits in the continuous improvement of the services for the stakeholders on which his or her business depends. Alas, the achievements of the responsible entrepreneur for society at large are neither part of daily media coverage on business and human rights issues, nor are they discussed in graduate courses in economics or business administration. What captures attention instead is the amoral, rational and utility-maximizing entrepreneur, who cheats and exploits where ever he can count on impunity in order to generate short-term profit. Even though these types of entrepreneur exist too, they usually generate too many enemies in business and society to survive in the long run. The publication of the 'Panama Papers' on April 3, 2016 and the 'Paradise Papers' in November 2017 illustrate well how easily abusive tax avoidance practices can be detected in the age of the digital society and lead to corporate reputation loss—even if the affected companies are able to prove that they did nothing illegal. Moreover, such media scandals result in blacklisting of states that are unwilling to adopt recommended OECD Transfer Pricing Guidelines for Multinational Companies and Tax Administrations,[16] incurring financial and reputation costs of states and corporations alike (Eggenberger 2018).

7.4.5 Chiquita: A Pioneer in Sustainable Banana Production with a Legacy Problem

Chiquita is a compelling case because of its long and controversial history of embeddedness. The history of United Fruit Company, as the company used to call itself before changing its name in the 1980s, was characterized by political rather than economic embeddedness. It played a crucial role during the Cold War period as a company that enjoyed the backing of the US government as well as the national land-owning elites in Latin American countries, which jointly fought against communism. As a consequence, the company's name became associated with client states in Latin America who were run by corrupt rulers that had more interest in serving US interests than their own people.

However, as Southgate and Roberts (2016) point out, the legacy of US-based banana companies in Latin America during the Cold War was not just about exploitation. Thanks to their commitment to embedding themselves into the local economy, United Fruit and other MNEs contributed to economic empowerment by teaming up with enterprising growers and exporters of national origins. This contribution was especially crucial in Ecuador, the world's largest tropical fruit exporter since 1953. The foreign investments in the local economy of Ecuador enabled the emergence of a powerful homegrown banana industry (Southgate and Roberts 2016).

Nevertheless, Chiquita was determined to leave the past of the Cold War behind as it embarked on a fundamental transformation of the company. In the early 1990s, it decided to change the way it was producing and exporting bananas radically. Its subsidiaries had to collaborate with Rainforest Alliance (RA) to improve their social and environmental sustainability performance in the early 1990s. In 2000 all Chiquita-owned farms were RA certified, and by 2012, the company managed to upgrade 75% of its local suppliers to become compatible with the RA standard.

---

[16]See http://www.oecd.org/ctp/oecd-transfer-pricing-guidelines-for-multinational-enterprises-and-tax-administrations-20769717.htm (Accessed Sept 25 2018).

In this period, its subsidiaries introduced a homeownership program for its employees. Local community projects such as "Nogal", a biodiversity reserve created in Costa Rica in partnership with Swiss retailer Migros, the Mesoamerican Reef Project in collaboration with WWF, and the Panama Biodiversity Project ("San San") with the support of the German retailer REWE, illustrate the company's commitment to the local environment. The company continuously improved the strict implementation of sustainability practices in banana cultivation and processing. It reduced its Carbon Footprint (in collaboration with the Massachusetts Institute of Technology, MIT) as well as its Water Footprint in collaboration with WWF International. These efforts are widely acknowledged by external reviews[17] and reported in the Economist (2012) under the title 'going bananas' because of the fact, that these corporate investments in sustainable local change failed to translate into an improved reputation. This because the company continues to be haunted by the past. Criticism always begins with what happened 40 years ago or more, whereas its transformation toward a global industry leader in sustainability remains unmentioned.

Chiquita's commitment to local embeddedness has been documented by three master theses at ETH Zurich, carried out in Guatemala, Costa Rica and Panama in 2016.[18] The critical field research conducted by the students discovered many positive side-effects for the local economy that resulted from the commitment of Chiquita subsidiaries to embed themselves into the domestic economy. Often these side-effects were not documented in Chiquita's sustainability report because they were attributed to the long-term business strategy rather than a concrete sustainability strategy. Even though Chiquita became a private company in 2016, after having been acquired by the Brazilian fruit juice producer Cutrale, the company continues to stick to its progressive CSR and sustainability strategy.

Many large retailers in Europe never acknowledged Chiquita's record as a pioneer in the field of CSR and sustainability in banana production, and those who did were unwilling to pay a premium price for their sustainable investments pointing out that consumers do not associate Chiquita with sustainability. Instead, many large European retailers prefer to exclusively sell fair trade and organic banana to their consumers to signal their sustainability commitment (Aerni 2013b).

Even Chiquita's long-time partner in the implementation of sustainability projects in Central America, the Swiss retailer Migros, decided to eventually delist its bananas in favour of Migros' own 'model project' of sustainable banana production in collaboration with WWF. There is, however, no evidence of any substantial sustainability improvements that would go beyond of what has been achieved in the prior collaboration with Chiquita (Schwab 2017). Migros is not the first retailer that literally 'bought' its WWF endorsement for its sustainable products in order to

[17]See Chiquita Assessment funded by the Fondation Guilé; see https://www.alexandria.unisg.ch/241457/1/Book-Interactif_300dpi.pdf (Accessed Sept 25 2018).

[18]26 See http://www.ccrs.uzh.ch/en/research/sustainable-impact/projekte/local-embeddedness.html (Accessed Sept 25 2018).

catch up in terms of sustainability reputation with competing retailers that were perceived to be more sustainable. Migros as well as Coop, the other dominant Swiss retailer, pay WWF on average around US$3–4 million per year to use their Panda bear. The German retailer EDEKA even disburses US$7 million in annual payments to WWF. Especially in the case of banana, WWF would have difficulties to defend such payments with the claim that their endorsement is justified by the added-value the business initiative creates for sustainability.

7.4.6 The Problem with Business to Consumer Labels in Agriculture

The claim that Business-to-Consumer (B2C) labels, such as fair trade and organic, would be per se the more sustainable alternatives to industrial agriculture is increasingly questioned. The so-called COSA study (2014) that compares the performance of different sustainability standards in agriculture suggests that the performance of organic and fair trade has not improved much, despite massive government subsidies, compared to less known Business to Business (B2B) labels such as 'utz' and 'Technoserve'.

'Jucker Farm', the most successful agritourism farm in Switzerland having won several national awards for sustainable business management and sustainable agricultural practices, refuses to label its agricultural products as 'organic'. This because it would require frequent spraying with natural pesticides, the use of heavy metals and liquid manure, something the majority of its customers and guests who frequent the farms' restaurant, use its recreational activities and purchase its homemade products, would not tolerate. Moreover, Jucker farm criticizes organic certification as being costly, bureaucratic and inflexible. It would not acknowledge that farmers may have a genuine interest in protecting and sustainably manage their natural resources, especially if their business model depends upon it. Their local knowledge and expertise allows tailoring sustainable solutions for their particular circumstances.[19]

The challenges of organic farming labels in developing countries are also documented in numerous publications (Dragusanu et al. 2014; Ronald and Adamchak 2008; Paarlberg and Paarlberg 2013).

What is especially odd in organic banana and cotton production is that water is considered a natural ingredient and therefore not subject to any use restrictions. This makes it possible to produce organic banana and cotton with intensive irrigation systems in arid zones where pest and disease infestation rates are relatively low, but water is very scarce.

Certified organic and fair trade in the tropics may also not be conducive to embeddedness. Clark and Hussey (2016) infer that describing fair trade as an effort

[19]See http://www.juckerfarm.ch/farmticker/detail/beitrag/ist-bio-wirklich-besser-teil-1/ (Accessed Sept 25 2018).

to re-embed agricultural production into the local economy, in the way Polanyi envisioned it, is misleading. Considering the fact that fair trade standards are defined by stakeholders abroad, they tend to discourage embeddedness in the local economy. Moreover, they tend to discourage innovation and entrepreneurship because the premium for members of the fair trade cooperative does not create any incentives to change or improve practices or search for new business models. In this context, the effect of the premium price is very similar to the effect of direct payments: they tend to cement existing structures rather than enable sustainable structural change. The failure to integrate fair trade and organic farms that produce for retailers abroad into national sustainability strategies has become apparent in Costa Rica, a country which aspires to be the leader in sustainable banana production in tropical countries. The Corporación Bananera Nacional Corbana, S.A. (CORBANA), an organization partially owned by the Banana growers of Costa Rica, seeks to obtain the Geographical Indication (GI) accreditation of the 'Banano de Costa Rica' as a national sustainability label for Banana production.[20] As such it collaborates with all major stakeholders in the banana business and sustainability certification agencies. Oddly, fair trade and organic cooperatives are not part of this collaboration. They argue that they have to follow the instructions demanded by the retailers in importing countries.

7.5 Embeddedness and Its Link to Sustainability and Corporate Responsibility

The case studies in the previous sub-chapters show that the term 'embeddedness' has a strong relationship with corporate responsibility and sustainability. After all, a private organization that is locally embedded while simultaneously abiding by its self-imposed principles of corporate responsibility is also more likely to assume active responsibility for the local people and their environment while also bringing prosperity to the region.

The ISO 26000 Standard on Corporate Social Responsibility (CSR) points out that multinational corporations should get involved in local community development. 'Community Development and Involvement' is, in fact, one of the priority areas of ISO 26000 and can be interpreted as a commitment to local embeddedness. However, the wording of the ISO 26000 Standard very much reflects the spirit of the UN Guiding Principles on Business and Human Rights (UNGP) and its focus on minimizing potential human rights risks resulting from FDI. In accordance with the view of embedded liberalism and analogous to the UNGP (Ruggie 1982), the purpose of ISO 26000 is to tell MNEs how to become good corporate citizens by refraining from doing bad things to people and the environment. The defensive narrative is increasingly becoming a problem for the implementation of Agenda

[20]See http://www.wipo.int/ipadvantage/en/details.jsp?id=3111 (Accessed Sept 25 2018).

2030 designed to meet the ambitious Sustainable Development Goals (SDGs) by 2030. The main purpose of the Agenda is not to avoid technological and economic change but to harness it for sustainable development. In this context, embedded FDI is essential for the creation of a sustainable global partnerships (SDG 17). It provides a foundation for SDG 8 (inclusive growth), SDG 9 (inclusive and sustainable industrialization), SDG 10 (reducing inequality through the people's economic empowerment) and SDG 16 (promote peaceful and inclusive societies). Moreover, since embedded MNEs are likely to create more jobs and economic opportunities in the local private sector, the resulting increase in average local incomes automatically lifts people out of poverty (SDG 1) and improves access to basic human rights as the right to food (SDG 2), health (SDG 3), education (SDG 4), equal treatment of gender (SDG 5) and clean water (SDG 6).

Chapter 8
Development Cooperation as a Catalyst for Sustainable Long-Term FDI

In 2002, the Organisation for Economic Cooperation and Development (OECD) published a report called 'Foreign direct investment for development: maximising the benefits, minimising the costs' (OECD 2002). Even though the report does not address the issue of principled embeddedness, it illustrates the many benefits that result from FDI in developing countries. The OECD Committee on Investment and Multinational Enterprises (CIME), which was responsible for the report, pointed out that MNEs need a favourable national framework to have incentives to invest in low income countries. However, favourable conditions alone may still not induce an MNE to invest in a high-risk developing economy, if the risk of reputation loss is high due to the challenge of implementing a due diligence process that is capable of controlling and containing any action that runs counter to the UN Guiding Principles on Business and Human Rights (UNGP).

MNEs already operating in high-risk economies may reduce due diligence risks by replacing local suppliers that are difficult to monitor and unfamiliar with global environmental, social and government (ESG) standards, with companies in economically advanced countries that are already familiar and able to effectively implement the required standards. As a consequence, the MNE may become ever less embedded in the local economy and eventually turn into a mere enclave of the affluent world. The growing share of intra-firm trade in international trade, as revealed in official trade statistics (UNCTAD 2013), may be an indirect indicator of this trend.

Especially after the global financial crisis in 2008, the share of intra-firm trade started to grow compared to trade between independent trading partners (World Bank 2017). In 2015, intra-firm transactions accounted for about one-third of global exports (UNCTAD 2013). This sort of trade may be quite valuable insofar as it allows MNEs to internalise transactions if the cost of performing these through the market is higher than internal costs (Coase 1937). Even though most intra-firm trade probably does not prevent the integration of local companies into GVCs, there is a certain risk with buyer-driven B2C standards in the global retail business, since, as shown in Sect. 7.4.6, they tend to prevent integration of the local economy into global value chains (Aerni 2017c). The costs to comply with one-size-fits-all retail

P. Aerni, *Global Business in Local Culture*, SpringerBriefs in Economics,
https://doi.org/10.1007/978-3-030-03798-7_8

standards on 'good agricultural practices' are lower in low income countries, if the business is run exclusively by people from the country where the global retailer's headquarters are based. After all, they know better how to comply with the self-imposed rules. As a result, professional consulting services are imported to ensure compliance with UNGP on business and human rights on plantation sites in developing countries. This may benefit people employed in the sourcing country as a quasi off-shore production site of a prosperous economy in the North. It does not contribute much to homegrown economic development, which would be necessary for those people who are not employed by foreign companies, namely the self-employed, the so-called survival entrepreneurs. They are often worse off than the employed (Aerni 2015a). The OECD itself does not seem to be concerned about the increasing reluctance of subsidiaries of MNEs to integrate into the local economy due to possible uncontrollable reputation risks. The organisation implicitly assumes that the value of FDI for the domestic economy is the same, whether trade is intra-firm or at arm's length[1] (Lanz and Miroudot 2011).

In this context, the OECD guidelines for multinational enterprises on responsible business conduct confirm the indifference toward the need of FDI to become embedded in host countries. They contain the same bias toward risk avoidance and guardian morality at the expense of risk-taking and entrepreneurial morality like the UN Guiding Principles (UNGP) on Business and Human Rights; they also include an implementation mechanism in the form of National Contact Points (NCPs). The NCP's offer a forum for discussion and assist interested parties concerned with business and human rights issues by conducting publicly available inquiries. Even though NCPs have the potential to improve incentives for MNEs to act responsibly in developing countries, they may not encourage the embeddedness of FDI in high-risk developing economies.

A recent article by Dorobantu and Flemming (2017) in the Harvard Business Review confirms the view that FDI's emphasis on risk avoidance is primarily framed as an issue of reputation management. The authors explore the question why companies should listen to local communities and discuss the question how to manage the social risks of projects that require a big amount of capital investment. The authors recommend formal engagement with local stakeholders rather than economic embeddedness as an effective way to manage corporate reputation. The approach is guided by a risk aversion strategy that may lead to more intra-firm trade converting local subsidiaries essentially into offshore production sites that profess to listen to local stakeholders but mainly serve the needs of stakeholders elsewhere (Aerni 2017a).

Another OECD report on FDI, published in 2002, makes the critical point that Development Assistance (DA) should not be a substitute for lack of FDI but encourage more of it. The report illustrates this by showing how carefully targeted DA can assist MNEs in leveraging FDI flows and creating a virtuous circle of

[1]Whereas intra-firm trade consists of cross-border transactions between firms linked by a degree of control and ownership, arms-length trade is defined as mere cross-border transactions. Vertically integrated firms trade primarily intra-firm (World Bank 2017).

increasing savings and investment in the respective region of investment. In this context, OECD may have anticipated a trend that has gained prominence since the adoption of the UN SDGs in fall 2015.

8.1 How Development Assistance (DA) Could Encourage Foreign Direct Investment (FDI)

The view that DA can be used to buttress or mobilise sustainable FDI has encountered a growing interest with the increasing importance of the term 'impact investment'. Impact investment comprises a wide range of funding activities designed to generate a measurable, beneficial social and/or environmental impact alongside a financial return (Brugg-Levine and Emerson 2011). Whereas the 'measurable' or 'evidence-based' part of impact investment, based on clearly defined interventions, has become a focus of economic research and fashionable in DA, it has delivered very few tangible results in terms of long-term development and economic empowerment (Aerni 2015b). It may be because most intervention studies are conducted while a certain project is still being funded. The primary challenge is, however, related to the fact that any externally funded project designed to improve human health, education or the environment in low income countries is likely to be discontinued if local people see no opportunity to generate an income from it—beyond what is made available by the external donor (Aerni 2015b). Moreover, evidence-based studies treat local people as passive subjects rather than entrepreneurial and innovative actors that change their economic, natural and social environment in unexpected ways (Olpadwala 2012).

There is however one branch of impact investment that is less focused on immediate results funded and monitored by a particular private or public institution, but concerned with the ability of FDI to contribute to inclusive growth through the creation of a new local business ecosystem that is financially sustainable and economically scalable (Frei and Schluep 2017).

A recent publication by Kaplan et al. (2018) outlines the philosophy behind this new concept of CSR: It suggests that subsidiaries of MNEs, which aim to integrate local entrepreneurs and skilled workers into the formal economy, must first understand and reimagine the regional economic ecosystem. In this context, they have to establish local relationships with stakeholders who are economically relevant, but may have little knowledge about the business at stake. In this context, a subsidiary of an MNE should not regard itself as a passive enforcer of a global corporate strategy, defined in the headquarters elsewhere, with a separate budget to fund some local community projects meant as a CSR contribution. Instead, it should discuss with local stakeholders the factors that prevent the MNE from investing in the local economy and its people. For that purpose, there is a need for a trusted facilitator, be it a local NGO, consultancy or government agency, that preferably obtains financial support from the DA community. It brings together the stakeholders that are prepared to discuss possibilities to take advantage of the presence of the foreign company by building a local economic ecosystem around it, which is able to reduce the risks of the MNE subsidiary to source things and sell things locally. DA that invests in such agents of change help to crowd in FDI and encourage incentives to become locally embedded, whereas the prevailing classic DA tends to crowd out FDI by building up parallel economic structures, e.g. through heavily subsidised alternative livelihood projects that mainly reflect the priorities of the donor country.

8.1.1 Sourcing Ingredients for Beer Production in Uganda

The Uganda Value-Added Maize Alliance (UVAMA) was a highly successful effort to integrate semi-subsistence corn-growing farmers into formal international value chains that use corn as a major ingredient in their production process. The catalysts who enabled this integration of the local private sector into GVCs was a global economic development consultancy called Carana (now owned by the consulting firm 'Palladium') in collaboration with AgroWays Ltd., a local grain trading company.

USAID, the agency for development assistance of the United States, provided the necessary seed money in the form of matching grants for the creation of a local formal supply chain that eventually succeeded in inducing MNEs based in Uganda to source some of their agricultural ingredients locally rather than importing them from elsewhere (Kaplan et al. 2018). One of these MNEs is the company Nile Breweries Limited (NBL), owned by SAB Miller (now part of Anheuser-Busch InBev). It used to import most of its ingredients for beer production from abroad to produce, sell and export its beer in Uganda. The Ugandan government aimed earlier at discouraging the preference of foreign companies to import their ingredients by encouraging instead domestic sourcing through tax incentives. Initial trials to buy maize and maize grit grown locally were however unsuccessful because local grain traders were unable to provide the required quantities in the necessary quality.

UVAMA was different. In a first phase, the Market Linkage Initiative was launched in collaboration with AgroWays Ltd. It was designed to improve quality, reduce post-harvest losses and invest in farmer's ability to comply with NBL procurement requirements. Among other things, investment went into the creation of eight Village Aggregation Centers (VACs) and a Grain Bulking Facility (GBF) with advanced post-harvest technologies that ensured the necessary grain quality. Because the Uganda Commodity Exchange licensed AgroWays, it could also offer farmers access to credit through a warehouse receipts program at its GBF. This greatly enhanced the number of farmers that were able to invest in making their farms more productive. This successful two-year pilot project eventually induced NBL to participate in the second phase with another three-year USAID matching grant for the actual creation of UVAMA. UVAMA allowed AgroWays to invest, among other things, in modern equipment that enabled it to process maize into grits. It finally induced NBL to source this ingredient locally, adding significant value to the local economy and improving the income as well as the agricultural practices of local farmers.

It ultimately helped local farmers and many other local companies involved in the production, sourcing and processing of maize ingredients to improve the quality and quantity of corn production and its derived products. Annual farm household revenues nearly doubled, and 70% of the sales of AgroWays in 2016 went to NBL (Tugendhat 2017).

8.1.2 Empowering Pastoralists Through Business Development in Kenya

In 1981, 16 members of a Keekenyokie pastoral community, which belongs to the ethnic group of the Maasai, decided to establish an abattoir in Kiserian, Kajiado County, Kenya, 40 km south of the capital city of Nairobi. The purpose was to sell their processed cattle meat to retailers in the capital and, with it, to increase household revenues for the pastoral community. They eventually registered their business as a formal company under the name of Keekonyokie Butchers Limited in 1992. The business operates in a large and complex market system in which information on price and compliance with quality standards is crucial. To cope with this complexity, the Maasai-owned company eventually started to make use of advanced information and communication technologies, offered by external service providers, to facilitate better access to market information and ensure the security of its transactions. The external service providers were local private companies that benefited from upgrading through public-private partnerships that involved international DA as well as foreign companies with particular expertise. They helped to link the different value chain players, including those in the informal sector, to markets, financial services and market information. As such, they ensured compliance with deliveries of agreed volumes and qualities of the respective produce at an agreed time and place (Nyende 2011). As a consequence of this upgrading and professional integration into a complex value chain, the Maasai-owned slaughterhouse scaled-up its meat production and processing capacity substantially. It had increased its processing capacity from 30 cattle a day in the 1980s to around 180 a day in 2010. The daily turnover from the slaughtering, resale and re-stocking of cows and goats in Kiserian has reached around US$60,000, and the slaughterhouse alone employs 176 people (FAO 2013). It eventually made the Keekonyokie livestock market the largest pastoralist owned livestock marketing and processing enterprise in East Africa. It has provided improved livelihoods to more than 100,000 households in Kenya and Northern Tanzania.

However, the growing business also led to a severe problem with the discharge of waste into the environment. In 2002, the company was asked by the National Environmental Management Authority (NEMA) to pay for the clean up of the waste resulting from animal slaughtering and processing. This was very costly and made the venture almost go bankrupt. However, at that time, the German GiZ (Gruppe für

Entwicklung und Zusammenarbeit) started to become interested in promoting sustainable energy solutions based on the use of organic waste. It initiated a public-private partnership (PP) designed to install and operate biogas production plants in rural areas as a source of energy (AHK Kenia 2014). Especially the waste produced by abattoirs was considered to have a high potential of biogas production at competitive prices.

In 2005, this Maasai-owned slaughterhouse company started to collaborate with GiZ to construct and operate a biogas plant. In 2011, the company presented a container prototype for the packaging of biogas intended for sale to small local private households; in addition to its existing sales of larger quantities of biogas to local institutions such as hotels and hospitals, and as the leading energy provider of the slaughterhouse itself. The sale of secondary products such as biogas to local households and institutions offered a competitive substitute to the primary source of fuel in the region, which is wood. After all, its zero-cost raw material comes from an average of 120 cows and 400 sheep and goats slaughtered daily (Karuga 2015).

The case of the Keekonyokie Slaughter House venture illustrates how local traditional entrepreneurs can be assisted by intermediaries that link DA to private sector investment in efforts to scale up production and capture additional business opportunities by making use of secondary products. The case study also illustrates the tremendous external social and environmental benefits resulting from a professional and profitable business run and owned by the locals. As for the economic and environmental benefits resulting from the construction of a biogas plant, GiZ played the catalyst that mobilised the know-how and the technologies on biogas production in the private sector and initially assisted in the funding and operation and maintenance of the plant. Since this secondary business eventually also became profitable for the local Maasai, GiZ was able to make a successful exit.

8.1.3 Enabling Access to Finance: The Case of Vodafone

Much has been written about the success of M-PESA, a mobile money transfer service launched in Kenya in 2007. M-PESA facilitates deposit and withdrawal of money through authorised agents—usually small mobile phone store owners—who convert virtual money to cash. M-PESA is also used to pay salaries, school fees and even traffic fine payments. A vast number of shops also accepts it as a means of payment. The transfers are safe as they are secured by a PIN number and confirmation text messages to both the sender and the receiver.

In 2016, the service has reached more than 25 million registered customers and more than 100,000 authorised agents. Its mobile financial transactions, used in a wide variety of situations, had a very positive economic impact on the daily lives of millions of people across Africa (Suri and Jack 2016). It led to the financial inclusion of the unbanked resulting in an increase in local money circulation, money services and capital accumulation, which in turn led, directly or indirectly, to significant improvements in areas such as food security, business expansion and employment opportunities.

The idea to create M-PESA goes back to three employees of Vodafone UK: Nick Hughes, Michael Joseph, and Susie Lonie. Hughes and his team were in search of new business opportunities for their mobile phone company in Africa. When attending a conference on the promotion of financial services for development, Hughes noticed that no one spotted the potential of mobile phones for enhancing financial inclusion. He and his colleagues wanted to prove that there is a market to serve the unbanked in Africa thanks to the rapid adoption of cheap mobile phones in the informal economy. Kenya was identified as a first test market because of its high adoption rates of mobile phones and the fact that the dominant local mobile phone service provider Safaricom, was a subsidiary of Vodafone.

Vodafone UK was, however, reluctant to support their initiative. The business case was not yet convincing, and a venture with an untested idea was considered too risky. In the search for alternative sources of funding Hughes and his associates were able to obtain an initial grant of one million UK£ from the Department for International Development (DFID), the UK aid agency. The initial purpose was to create a platform to use mobile phones for microfinance repayments. However, the pilot project with 600 local participants quickly showed, that Kenyans were more interested in making use of their mobile phones as a tool for financial transactions rather than for microcredit management. The platform was adjusted accordingly, and the Kenyan government eventually became convinced that Safaricom was sufficiently large and professional to guarantee transaction security. Twelve months after its launch in 2007, more than 2 million people were already using the service. Once it became clear that there is a genuine business case, Vodafone invested in the new market on a large scale. Today 17 million Kenyans use the service, and Safaricom is expected to make a record profit in 2017 of US$500 million. At the same time, the venture proved to be one of the most successful initiatives to reduce poverty through the economic empowerment of the poor, and women in particular (Urech 2018).

The M-PESA success story also laid the groundwork for other businesses to be developed in Kenya, such as M-Kopa solar (a brainchild of Jesse Moore and, once again, Nick Hughes), a 'pay-as-you-go' provider of solar energy systems to off-grid homes in Kenya. In 2018, M-KOPA has connected 600,000 homes across Africa. It provides households with 75 million hours of kerosene-free lighting each month. The customer package does not just contain a battery-powered system but lights, phone-charging, and a solar-powered radio. All combined including daily mobile money payments, costs less than the cost of traditional fuels.

Since 3G internet connections are increasingly affordable, a promising startup scene and ecosystem has evolved in Kenya (also known as Silicon Savannah) that is primarily focused on adding new mobile phone services that respond to local needs. As uptake and affordability have increased M-Kopa is already facing stiff competition from other providers such as Off-Grid Electric, Azuri Technologies, and Mobisol, all of whom are targeting the same low-income customers with payment plans that amount to less than the aggregated daily expenses for kerosene (Dahir 2018).

8.2 Lessons Learned from Public-Private Partnerships (PPPs)

The previous examples show how investments by MNEs in developing countries can generate significant social benefits provided that government, as well as civil society organisations, assist them in their efforts to become embedded in the local economy. Unlike the social benefits resulting from small CSR projects, the social benefits that result from profitable businesses may be more relevant to poverty reduction since they are scalable. In other words, they are profitable and therefore financially sustainable, which makes possible a successful exit of the initial catalyst, be it a company or an aid agency. The business is then owned by a local entrepreneur, who aims to keep the company's lead in the new market, despite growing competition from new players who emerge once it becomes known that the venture proves profitable. In other words, the initial venture becomes a prospering and innovation-driven domestic market that generates employment, provides essential goods and services and adds to the local tax base. All this contributes to long-term economic empowerment and inclusive growth.

The examples also illustrate that an MNE that is under pressure to deliver return on investment, faces the problem of small volume and high uncertainty. It may therefore be unable to invest on its own in a local market—unless it can count on highly professional local partners that help address the essential coordination challenges. Preferably, they are not under pressure to deliver short-term financial results and have the necessary authority, legitimacy, and trust in the region to mediate between the needs of the MNE and the capabilities and resources in the local private sector. As such they address all the organisational and cultural hurdles that make socially inclusive business models and business ecosystems so challenging to implement and scale up.

Moreover, these local intermediaries with their mission to create inclusive growth ecosystems may need the support of DA for testing their proof of concept in pilot projects. Once a pilot is successfully launched with seed funding and has reached the 'proof of concept' stage, the risk for the foreign company to invest in the commercialisation of the product decreases rapidly. Funding from DA may also create leverage to mobilise additional funds necessary to scale up the project rapidly. Eventually, the anchor corporation may become the principal investor, since funding to scale up an existing ecosystem is perceived as less risky and more likely to generate returns in the long run.

With growing experience and an increasingly sophisticated industry specialized in solving coordination challenges, the integration of local economies into global value chains will become less costly and less uncertain. This may then also render impact investment funds more attractive because the ventures are more likely to generate sustainable returns. Impact investment funds currently manage about $80 billion in assets worldwide.[2] It indicates that investments in sustainable change may eventually become mainstream in the financial sector, thanks to the fact that the public sector initially provided the project funding for the pilot trial necessary to jumpstart the catalytic process of local economic development.

[2]See http://www.undp.org/content/sdfinance/en/home/solutions/impact-investment.html.

Chapter 9
Concluding Remarks

Whenever the mass media cover the role of multinational enterprises (MNEs) in developing countries, the storyline is nearly always the same: outsourcing is profitable if labour costs are low. Cheap labour is the reason why the textile industry moved from Europe to Asia and, more recently, has set up shop in Africa. Meagre salaries in the countries in the Global South seem to confirm the widespread view that MNEs are making profits by increasing human misery. Reading about the wages of workers in the textile industry that barely allow them to make a living causes a feeling of indignation, especially when then looking at the price tags in the urban department stores next door where a shirt often costs less than a combo sandwich.

If one then reads Karl Polanyi's 'Great Transformation', everything seems to make sense: there is the global expansion of formal markets dominated by MNEs that only care about shareholder value and the evasion of taxes at the expense of the precarious lives of local stakeholders. Local workers have been lured into abandoning their culturally embedded traditional economic systems in the hope of earning a better life in the global economic system. Yet, what they find is a system of exploitation that leads to cultural disembeddedness without any real gains in living standards.

The content of this book has shown that the situation is not as clear-cut as it seems. Many MNEs that are committed to embedding themselves in the region in which they invest may add significant value to the local economy. This is illustrated in Chap. 7 using four case studies that cover prominent MNEs: Nestlé, Syngenta, Bata Shoes and Chiquita. Rather than indiscriminately devaluing local cultures, the subsidiaries of these companies in developing countries tend to contribute to new hybrid forms in which global elements do not replace but get integrated into the local culture.

At the same time, these MNEs are committed to complying with their internal codes of conduct to avoid negative forms of embeddedness associated with collusion and corruption. They practice a policy of transparency and openness and follow industry-wide social and environmental standards. In order to assess the

P. Aerni, *Global Business in Local Culture*, SpringerBriefs in Economics,
https://doi.org/10.1007/978-3-030-03798-7_9

overall environmental and socioeconomic footprints of these foreign investors, the positive impact of this type of 'principled embeddedness' must be balanced against the negative impact related to the ecological footprint of the core business and its extraction of local resources, scandals related to child labour, tax evasion, treatment of employees and so on. MNEs with a positive net impact and a commitment to the region in which they invest, may therefore not be part of the problem but actually be part of the solution when addressing local sustainability challenges and contributing to inclusive growth.

Yet, there are also negative examples of MNEs in the retail and fashion industry that may invest a lot in their reputation as sustainable companies but refrain from becoming locally embedded through a genuine long-term engagement with local stakeholders. They tend to impose instead their self-defined one-size-fits-all sustainability standards, which allow them to dispense with local interaction and instead delegate the auditing to professional international third parties. Such standards usually do not contain any local input. As such they essentially replace, and with it, devalue local knowledge that may be highly relevant in the local context (see Sects. 1.3, 4.2, 7.4.6 and Chap. 8). Since every supplier has to comply with the same buyer-driven standards, independent of location and stage of the development of the respective region, it makes sense that the suppliers are either 100% controlled by the foreign company or that the foreign company delegates the responsibility to a profit-oriented international third-party certification agency. The certification agency has to ensure that local suppliers comply with the required standards (while bearing the costs of compliance themselves). They inspect the sites and report back to the contractor. In other words, communication with the local suppliers is essentially reduced to a global audit culture based on compliance procedures defined and enforced by external stakeholders (Merry and Conley 2011; Power 1997). This can have severe consequences for the local suppliers and their employees who have been virtually cut out of corporate communication.

The collapse of the Rana Plaza garment manufacturing building in Bangladesh in May 2013 illustrates the consequences of relying merely on external auditing. The accident killed more than a thousand people. All these lost lives could have been prevented if the global fashion retail companies had listened to the warnings of the local textile workers who found cracks in the poorly constructed building, which made them scared to go to work. Yet, the sustainability standards of the global fashion brands, such as the Business Social Compliance Initiative (BSCI), have been designed elsewhere in collaboration with international NGOs. Since the safety of building was not on the radar screen and therefore not included among the factors auditors were asked to consider, it did not exist officially as a sustainability issue.[1] Local workers may report problems, but if they are unrelated to any particular paragraph in the labour standards, the reporting may not be heeded.

[1]See for more details: https://cleanclothes.org/news/2013/06/25/bsci-10th-anniversary-shame-over-rana-plaza (Accessed Sept 25 2018).

This example illustrates well how conventional corporate social responsibility (CSR) initiatives have created a corporate culture that is not assuming active responsibility but instead merely delegates it. The focus is on damage control in the communication with stakeholders in high-, rather than low-income countries. The results are standards that reflect the concerns of shareholders, affluent consumers and global activists, but not necessarily those of local stakeholders in producer countries (Aerni 2013a; Freidberg 2007).

All of this illustrates the pitfalls of a disembedded global economy replacing informal local interaction with standard procedures of communication following the principle of measurement, reporting and verification (MRV). This codified, universal knowledge that the Greeks called episteme has become a mere risk management tool that allows global business to manage issues related to local context with minimal communication with local stakeholders.

At least one could argue that the textile manufacturing industry may have learned its lessons and would care more about local context (Schuessler et al. 2018). But, as illustrated in Sect. 7.4.6, the global retail industry, as a whole, did probably not. It continues to stick to the same one-size-fits-all sustainability standards that often proved to be inappropriate because they ignore local context. This applies, in particular, to Business-to-Consumer (B2C) standards, such as fair trade and organic. These labels may have contributed to environmental improvements and social justice in the early 1980s when fair trade and organic cooperatives were still grassroots movements that tried out alternative forms of sustainable agriculture. But with the codification, commodification and standardisation of these premium labels by global retailers, the grassroots character disappeared and gave way to a global auditing business with a hardnosed tick-box mentality (Freidberg 2007).

This global auditing culture has led to the absurd situation, that B2C premium labels, such as fair trade and organic, are often underperforming in low income countries, compared to Business to Business (B2B) labels that the agribusiness and textile manufacturers have designed in collaboration with local value chain partners to improve their sustainability performance (COSA 2014; Lott 2015; Haight 2011). This trend makes sense since there is little incentive to improve further when retailer marketing already does the job of praising your label as the only moral and sustainable choice. The premium price that consumers are willing to pay for the warm glow of the 'ethical' label ensures a high-profit margin and a boost in reputation for the retailer. If consumers are happy and margins are high, why raise critical questions and ask for change?[2]

This self-righteousness has however frustrated the supplying industry, which invested a great deal in making their global value chains (GVC) more sustainable. Retailers proved unwilling to acknowledge these efforts by moving beyond the low

[2]There are certain initiatives, such as the Iseal Alliance (https://www.isealalliance.org/) that claim to recognise the fact that sustainability standards need to improve and become more inclusive, based on mutual learning and involving all stakeholders in the value chain. But the alliance is again funded by the same agencies that promote a defensive understanding of sustainability (retailers, organisations involved in DA).

wholesale price for products that may perform better in terms of sustainability but lack the 'goodness' reputation of well-known B2C labels such as fair trade and organic (Economist 2012).

Another problem with B2C labels is that they have created economic enclaves of European retail business in low income countries. These off-shore islands of 'sustainable retail business' are farm cooperatives that sell their certified fair trade and organic products exclusively to the international retailer, who then sells it again for a premium price to affluent consumers elsewhere. The farmers in the cooperatives also benefit from the premium price, but at the expense of becoming a 100% dependent on external players. They get all the instructions on what to produce, how to produce and when to deliver from abroad. This arrangement reminds one of a benign form of the master-servant mode of pre-industrial Europe with its feudal overlords, as described in Chap. 3. But as the lives of servants' that operate in the modern fair trade and organic cooperatives seem somewhat privileged compared to their local peers who produce for the local market, they have no interest in changing the convenient arrangement or take the risk to become local entrepreneurs responding to needs and opportunities in the regional economy. There is also little encouragement from the side of their foreign-sponsors to use the higher revenues, stemming from premium prices, for business ventures that help address the local needs and diversify economic activities.

So even if these cooperatives operate in a fair and sustainable way, the question remains whether they contribute to the UN Sustainable Development Goals (SDGs)? The enormous potential of the private sector to contribute to the SDGs is seen in inclusive growth (Goals 8), infrastructure and innovation (Goal 9) as well as less inequality through economic empowerment (Goal 10). Since retail-sponsored fair trade and organic cooperatives in developing countries represent mere niche markets for affluent consumers elsewhere, they contribute little to nothing to these goals, mainly because local embeddedness is not one of their concerns.

In this sense, Polanyi may be right about a trend of the global economy to disembed local culture. But ironically, this trend is not due to MNEs in agribusiness but instead caused by global retailers, civil society organisations, and government agencies involved in DA and trade promotion. These actors share a defensive understanding of sustainability, focused mainly on preserving and protecting people and their environment from the forces of global economic change. This goes well together with the goals of externally sponsored fair trade and organic cooperatives that shield the members off from the low-wage and risky local economy that surrounds them. The surrounding low-wage economy is however present within the well-protected cooperatives in the form of migrant workers, who assist members of the cooperatives during labour-intensive periods. As non-members, they do not have the privileges of prime access to health, education and above-average income (Haight 2011).

At any rate, retailers generously remunerate environmental organisations that are known to be popular with affluent consumers for their endorsement of fair trade, organic and other consumer products with B2C labels that signal 'sustainability'. In the case of WWF, this remuneration has reached levels that may render the organisation financially dependent on the retail industry while the contribution of these products to the SDGs, is often questionable (Schwab 2017; Aerni 2017b). In addition, government agencies involved in DA have also joined the bandwagon, especially when it comes to the disbursement of 'aid for trade' money. As illustrated in Sect. 6.5, OECD countries have made generous use of the means available in DA for sustainable trade to assist their retailers with global reach in the set up of fair trade cooperatives in developing countries; officially portrayed as a contribution to 'aid for trade', allegedly responding to priorities in low income countries.

Polanyi's bipolar framing of 'global profits versus the local people' might make much sense in the context of the promotion of 'sustainable' niche markets designed to enhance the reputation of global retailers and donors in affluent countries. But how is it possible, then, that all the global activists and academics who celebrate Polanyi's attack against the capitalist system in 'the Great Transformation' remain silent about these almost neocolonial patterns emerging in the retail and the aid industry?

This silence may be related to the match between the myth of organic and fair trade and the popular defensive understanding of sustainability in postmaterialist societies. It also fits the implicit bipolar baseline assumption that economic growth merely generates social costs that subsequently need to be mitigated by the welfare state. Chap. 3 points out that this view is entrenched in the neoclassical paradigm in welfare economics as well as in structuralist theories in sociology, anthropology and human geography. These theories have their roots in the Cold War period but continue to prevail at universities in the 21st century, sometimes repackaged under different labels such as environmental economics or constructivism. What they all share is a disinterest in the historical dimension of human interaction and a belief in the existence of universally applicable laws and structures of social and economic exchange, independent of time and context. The Chaps. 2 and 4.2 illustrate the influence of Polanyi on these enduring schools of thought represented by luminaries such as John Ruggie, a representative of the school of embedded liberalism, Dani Rodrik, author of the book 'The Globalization Paradox' and Former World Bank Chief Economist and Noble Prize Winner Joseph Stiglitz.

The implicit view that 'the global' undermines 'the local' also explains the sympathy that these well-known academics have for anti-globalisation activists who claim to defend local culture against global business while having become a product of global culture themselves. Section 4.3 illustrates this in the case of Vandana Shiva who gained her credentials as a global activist by popularising the Indian Chipko movement as the tree-hugging movement in the United States. Later she became well-known as a critic of the green revolution and the use of genetic engineering in agriculture. Her claims are often based on anecdotes that have either not been validated, or if validated, turned out to have completely misrepresented the issue at stake. In the case of the Indian Chipko movement, she portrayed it to the Western audience as a symbol of ecofeminism because local women were hugging trees to prevent them from being felled by logging companies. However, the Chipko was a very heterogeneous movement, united by locals who fought for their economic right to make commercial use of their forest resources. Shiva knew, however, that such a story would not appeal to postmaterial affluent societies and brought it more in line with the prevailing bipolar mindset, hailing the local and cursing the global. Her mythical account of how the local Chipko activists successfully defended themselves against external corporate invaders and eventually restored the original environmentally and socially sustainable livelihoods sounds very plausible to people concerned about globalisation. It contributes to an overall belief system that provides certainty, meaning, identity and orientation in a complex and demystified world. She then repeated the same feat with the story about Indian farmers committing suicide as a result of the adoption of Bt cotton. Empirical research showed however over and over again that small-scale farmers in India who adopted Bt cotton fared much better than their peers who continue to use lots of pesticides in conventional cotton production. This research had, however, no effect on the persistent narrative about the 'seeds of death' created by Shiva and her followers (Herring 2010).

As a global 'epistemic brokers', Vandana Shiva and other global activists have enormous influence on the way retailers design their normative marketing campaigns. In such campaigns, retailers do no more suggest to consumers what they 'can' buy but instead what they 'should' buy to save the planet, for example by avoiding 'industrial agriculture'. In other words, they claim to help consumers to make a moral choice between the 'the local' associated with 'the natural' and 'the good'; and 'the global', associated with opposite attributes. What is concealed in this distinction, is the fact that there is always a lot of 'global' in the 'local'. But who wants to know that really, if the choice also comes with a price premium that increases self-esteem in consumers and ensures higher margins for the retailer?

In addition to this commercial exploitation, the influence of the global campaigns of these activists on the formation of public opinion also allows them to shape not merely policy debates and the allocation of aid money, but indirectly frame research questions and set research funding priorities in social and environmental research as well. Section 6.5.1 points out how grant applications submitted in the area of research for development (r4d) in Switzerland must be primarily based on a defensive understanding of sustainable development in order to get funding. Most of the projects funded address topics such as environmental monitoring, climate change modelling, gender, organic farming as well as 'resilience' and 'vulnerability' of local communities. In turn, no research proposals have been approved that explore the contribution of public-private partnerships (PPPs) or technology transfer to inclusive growth; or the role of innovation and entrepreneurship for sustainable development. This is odd in view of the fact that the latter topics are actually on the priority list of the Agenda 2030 to implement the Sustainable Development Goals (SDGs).

The power of global influencers and anti-globalisation activists in shaping the sustainability agenda has caused a shift in the public perception of 'expertise'. While the science-based advice of expert networks used to frame problems and formulate the policies of decision makers to address these problems, this task is increasingly assumed by those who have shaped the popular narratives about the 'bad' global economic players who threaten the 'good' local stakeholders.

Science-based expert systems came under criticism decades ago for being too technocratic and disconnected from people's concerns (Rowe and Wright 2001). Instead global activists may have become the new experts. They claim to represent people's concerns, and the mass media has portrayed them accordingly as advocates of the public interest, and sometimes even 'independent science'. After all, media-relevant events need to be presented as a sort of drama with perpetrators, victims and heroes that come to rescue, and global activists can provide for it (Nelkin 1995). As epistemic brokers, they have accumulated sufficient cultural capital to create, via social media, a global community of believers in their political and scientific claims. Based on the popular anti-globalisation narrative, they create media- relevant dramas with a moral undertone, denouncing the supposed perpetrators, and speaking in defence of the victims. As such, they have gained public trust, which has enabled them to shape and control knowledge and information that is identified by the public as objective and legitimate. Global activists may claim to be 'opposed' to the 'system'; yet, they have gained a position of power within this very system that is comparable to the former power of technical and scientific expert networks. The rise of social media has accelerated the diffusion of their narratives and their claims, to such an extent, that they create new patterns of beliefs and frames of interpretation on a global scale that remain largely unchallenged (Nichols 2017).

Their targets of protest are not global retailers and how they promote a disembedded global economy based on tick-box standards, but actually those MNEs that are genuinely committed to embed their investments into local business and society; often because it also makes business sense to do so.

These MNEs operate in global mass markets in which labour costs indeed matter. However, if they strictly abide by industry-wide labour and environmental standards and are committed to continuous improvement through innovation and improved performance, they may contribute to economic empowerment, inclusive growth and sustainable change through a strategy of 'principled embeddedness' as described in detail in Chap. 1 and illustrated in Chap. 7. Yet, their efforts hardly ever get any recognition from social activists because it would make things complicated to shift from an established risk debate to a more balanced debate over risks and benefits of economic globalisation.

For MNEs, it may make sense to become more embedded, especially if they have invested a great deal in production facilities, and if they need to build up social capital in the host country to secure their license to operate. In other words, unlike retailers which focus on pleasing the affluent consumers in high income countries, MNEs that invest in production facilities in low income countries need to please the local stakeholders, as much as the stakeholders and shareholders abroad.

However, it was pointed out in Chap. 8 that not all industries have incentives to become more embedded in the local economy. These incentives diminish as uncertainty regarding the transactions costs involved in dealing with local business grows, and as sourcing inputs from abroad becomes easier and cheaper. The transaction costs of partnering with local business are particularly high if there is a lack of reliable physical and digital infrastructure, if young people with the skills that formal businesses require are scarce, and if there is a lack of local suppliers to provide the product and service quality that is required in the respective global value chains (GVC). MNEs alone are not able to address the countrywide challenges to integrate more informal businesses into the formal economy. There is a need, instead, for organisations that function as catalysts connecting local business to global companies by funding initial pilot projects to test if a partnership could be of mutual economic benefit and therefore eventually be scaled up. Concrete examples of such catalysts always involve progressive-minded organisations in development assistance (DA) committed to cooperation rather than confrontation. In this context, Chap. 8 discusses how DA and trade promotion could be wholly reinvented by focusing the funding not just on the work of catalysts but also companies that are able to make money as professional coordinators who build the necessary bridges between global business and the local economy and thereby facilitate inclusive growth. Alas, the media coverage on the role of MNEs in low income countries is still one-sided, especially when they operate in labour-intensive industries. Unbalanced media coverage also explains why state and non-state actors involved in DA refrain from collaborating with MNEs, not because it would not result in substantial benefits for the local population, but because it is seen as a possible risk of reputation loss for the state or non-state actor to have an MNE as a partner in DA. Sometimes, international financial institutions and global donor agencies that support public-private partnerships and encourage rather than discourage investments of MNEs in developing countries are even denounced as agents of 'neoliberalism'.

Chapter 3 illustrated how the term 'neoliberalism' is used and abused in the public discourse on globalisation to an extent that it has lost its meaning. It may have originally referred to the so-called Washington Consensus, a series of prescriptive policies for debt-ridden developing countries promoted by the International Monetary Fund (IMF), the World Bank and the US Department of the Treasury in the 1980s and 1990s. Relying on the principles of neoclassical economics, the representatives of these institutions argued, among other things, that fiscal discipline and the privatisation of state enterprises must be part of any economic reform package to get rid of debt and embark on economic growth. With the economic rise of Asia and the creation of Asian-based financial institutions in development funding, this consensus has become obsolete and the crude measures applied back then to restructure developing economies find little support today.

The mindsets of supporters of neoliberal policies, as well as their radical opponents, have both been falsified by historical research, especially by the representatives of the French school of historians called Ecole des Annales who have examined the impact of economic change on people's daily lives through archival research. These historians re-visited Karl Polanyi's claims about the fatal social consequences of the enclosure movement in the UK for rural areas before industrialisation (Sect. 3.1) and the harmful effects of international agricultural trade on hunger and starvation (Sect. 3.2). They were unable to confirm them. Contemporary field research such as the one conducted by Rangan on the Chipko movement in India (Sect. 4.3.1) mostly confirms the insights gained from the previous archival research. Sadly, field research and the mixed method approach of combining qualitative with quantitative approaches from different disciplines, as initially advocated by John Dewey (1905), faces a funding crisis due to a return to strictly disciplinary research and methodological purity in the social sciences that can be observed in the second decade of the 21st century (König and Gorman 2017). The practical impact of this research proved to be correspondingly low (König 2018).

A promising discipline that is still committed to field research and open to integrating findings from other social science disciplines is economic sociology. Economic sociologists have done seminal qualitative research on the embeddedness of economic relations in social relations, which can be combined productively with quantitative research in the fields of complexity science and political economy (Sect. 5.3). Economic sociology is also of high practical relevance, as outlined in Chap. 5, especically if it addresses the three major coordination problems that MNEs are confronted with when investing abroad. As such, the discipline does not simplistically contrast the global versus the local, but instead addresses the challenge of ensuring a thriving cultural and economic integration of the global into the local and reversed.

Research in the area of business studies has made use of the concept of 'embeddedness' to investigate the role of subsidiaries of MNEs in local economic growth, as discussed in Sect. 7.3. The insights gained from this research as well as the case studies discussed in Chap. 7 illustrate that FDI in developing countries does not have to lead to heteronomous economic structures, but may well contribute to economic empowerment and cultural renewal.

These insights are, however, hardly discussed in the CSR literature on business and human rights, as pointed out in Chaps. 7 and 8. The UN Guiding Principles on Business and Human Rights (UNGP) as well as the OECD Guidelines for Multinational Enterprises are strictly focused on 'do no harm'. As such, they do not take into account the potential positive external effects that MNEs may generate for the local economy and society.

What could be the alternative to the implicit bipolar baseline assumptions that prevail in CSR reports, in the public debate as well as in academic research? After all, the bipolar view of the global and the local is easy to communicate and helps to mobilise resources in defence of local traditions and practices against the encroachment of global business.

The answers to these questions may come from academic research disciplines in which field research is still emphasised. Also, investigative journalism may increasingly matter. Investigative journalists are increasingly questioning the extent to which the rhetoric and gesture politics of global activists against MNEs in developing countries corresponds to the experience and interests of the local people on the ground, whom activists claim to represent. These investigative journalists travel for several days to the respective locations to listen to different local stakeholders without making use of the convenient and almost mythical accounts of 'the local' versus 'the global' as a frame of interpretation. Their research reveals a more nuanced picture of the role of MNEs in developing countries, even when they operate in the industries often criticised for only caring about the local production costs, such as the textile manufacturing industry.

The journalist Angela Köckritz recently published an article in the German weekly 'Die Zeit' (Köckritz 2017) about the impact of the emerging textile industry in the Awassa Industrial Park in Ethiopia on local livelihoods. The content of this article is worth recounting in these concluding remarks because it highlights the value of FDI in low income countries, not from an academic, political or corporate perspective but from the perspective of the local community that is involved in the daily operations of this emerging manufacturing branch in Ethiopia. It is essentially about local materiality. Köckritz refuses to write about 'them' without bothering to be precise about who these local workers are. She spends one day and night with one of the local sewing ladies, a young woman called Tigist, 18 years old and single. Tigist earns 1055 Birr, equivalent to a little more than US$35 per month. She works eight hours per day, six days per week with no holidays and late shift. All this would be sufficient to write a story about the exploitation of women in low income countries by the global textile companies. But once the journalist asks Tigist if she is happy with her circumstances, she hears an unexpected story. Tigist explains that she grew up in a village 50 km north of Awassa. Without this job, her destiny would have been sealed a long time ago. According to rural tradition, parents pick husbands for their daughters once they have reached the age of 13. After marriage, their primary purpose is to bear children, serve their husband and hope not to die early because of birth complications or other health risks. Tigist did not want her life to be short and brutish and therefore escaped to the city of Awassa. A risky venture, yet one she thinks was worth it. Even with her meagre salary and heavy workload, yet she found an affordable room, which she shares with two other women working in the same factory. One has become her close friend. On weekends they go to church together and maybe take a short walk in the city. She explains in detail how she uses her salary: one fourth goes to rent and another fourth to food, clothing and other necessities. That leaves as much as half her salary (US $17) to save for unexpected expenses and for whatever she plans to do in the future. Tigist thinks that the company offers her a great opportunity to learn new skills, to get used to regular working hours and discipline and to get evaluated all the time, which triggers her ambition to improve continuously. She dreams of the time when she will have acquired sufficient skills and savings to buy a sewing machine and set up her sewing shop. She could then maybe get orders from the company that previously employed her, or maybe she could mend or even design garments and sell them to local people.

At first, the journalist is puzzled listening to this dreamer, but then she realises that this must be the first step of any empowerment process of women. She recognizes the empowerment potential that comes with a steady salary. A salary provides the young woman with economic rights and a sense of freedom. It further allows her to organise her life and set her own priorities. Last but not least, it causes her to envision a future that is wide and open. When talking to Tigist's supervisor, the journalist learns that the garment manufacturing company that sells to European fashion brands was founded in India and employs around 12,000 workers worldwide. The supervisor was born in poverty himself in Sri Lanka and found a job there in a textile factory, where he worked his way up to the level of a factory manager.

He then moved with the rising salaries first to Bangladesh and later to Ethiopia. He talks of the garment industry as a pressure industry. He is expected to double the size of the production in Awassa during the next eight years while keeping production costs at a level that ensures the competitiveness of his factory in the global market. He points out that workers are most productive when well-fed, healthy and well-trained. As such, he regards the sewing ladies as the main asset of his business.

Finally, the journalist interviews the chairman of the Ethiopian association of textile industries. He sees textile manufacturing as a critical industry that would enable the country to eventually build up its manufacturing industries. It reminds Köckritz of what she learned about the history of industrialisation in England in the 19th century. She also sees the contemporary risks for countries like Ethiopia; many other Asian and African countries started previously with a garment industry but failed to move on because of political turmoil, export-subsidies for the textile industry in high income countries that undermined the competitiveness of the domestic industry, and unexpected technological changes. During the 21st century, robots may become better at sewing and eventually make the job of the sewer redundant. For now, none of this affects Tigist's life in Awassa. Everyone is used to pressure, and she has learned to deal with uncertainty. If one thing is no more working, then one has no choice but to move on, never knowing whether destiny leads to prosperity or destitution. What is certain is that people are not passive victims but active entrepreneurs, by choice or by force. Global activists, scholars in academia, CSR managers and project leaders in DA remain mostly silent about this fact. Why? Because a large part of the legitimacy of their work is based on the assumption that the solutions need to come from them. Local people, in return, are merely seen as grateful recipients of largesse from wealthy nations. They are assumed to have no agency, as Rangan (2000) expresses it. As a consequence, their needs are often defined by external stakeholders starting from the assumption that they are exploited and uprooted by MNEs against their will, and that their only wish is to preserve their unique cultural habits and traditional economic activities.

Historians who investigated the daily lives of ordinary people during pre-industrial Europe debunked this view as fiction. The poor at that time were not particularly interested in preserving their cultural rights, but rather in making use of economic rights to change their material situation through entrepreneurship and innovation (discussed in Chap. 3). Even the German philosopher Hegel noticed that it is the servant class that drives change rather than the masters, because the master is happy with the status quo while the servant is not (Kojève 1975). So MNEs in developing countries may become the chief enablers of social mobility in these countries by providing economic opportunities where none have existed before. As such they may also contribute to the emergence of an economically empowered middle class that will eventually articulate its particular interests in politics.

In a way, the account of the young sewer in Awassa illustrates the human factor in the global capitalist system and illustrates how wrong it is to see the presence of MNEs in low income countries merely from a perspective of potential harm. As an investigative journalist, Köckritz does not judge the motives of the MNEs, but rightly assumes that they must be profit-driven in order to stay in business.

However, the profit-driven motive does not seem to undermine the potential positive external impact for local people who search for economic opportunities to increase their ability to make their own choices in life.

The journalistic insights are confirmed in the research of social historians and economic sociologists who understand that economic relations are always embedded in prior social relations. But, are the motives expressed by the young sewer to leave her village and start a new precarious but more self-determined life in the city heard and respected in public policy, in academia or civil society? As a survival entrepreneur who believes in her skills to change her economic situation, Tigist, the sewer, does not fit the stereotypical victim of neoliberalism, structural inequality or human rights offences. The same is true for the other examples illustrated in Sect. 8.1, such as Maasais who decided to create a formal company designed to process and sell meat to urban markets. They have realised that the only chance to ensure their cultural and economic self-determination, is to selectively participate in economic globalisation in order to generate their revenues and thus ensure that the culture remains attractive to the next generation. Even though there are efforts in DA to support and coach such entrepreneurial activities, they do not receive the attention they deserve.

In conclusion, economic globalisation need not undermine local culture if MNEs are willing to embed themselves into the local economy. As such, they contribute to a sort of hybrid development in which 'the local' is no more kept apart from 'the global'. It is partially driven by the requirements of FDI but also by the local people who are still free to decide what to adopt and what to reject from the realm of global business and culture.

Ultimately, it will be possible to achieve the ambitious SDGs by 2030 only, if we can move beyond stereotypical accounts of predatory MNEs that generate profits at the expense of local people and their culture. The resulting polarisation creates confrontation where there should actually be collaboration. Since the principle of cooperation is also one of the essential pillars of sustainable development, bi-polar mindsets are not conducive to achieving the UN SDGs.

What is required is a context-based and dynamic understanding of sustainable development. One that focuses more on the actual human dimension which drives the process of economic integration with all its risks and opportunities, rather than relying on mythical accounts that may be plausible and easy to recount but distort reality and ultimately lead to undesirable political, economic, social and environmental outcomes.

References

Abdelal, R., & Ruggie, J. G. (2009). The principles of embedded liberalism: Social legitimacy and global capitalism. In D. Moss & J. Cisternino (Eds.), *New perspectives on regulation* (pp. 151–162). Cambridge, MA: Tobin Project.
Acemoglu, D., & Robinson, J. A. (2013). *Why nations fail: The origins of power, prosperity and poverty*. Profile Books.
Aerni, P. (2017a). *Beyond corporate social responsibility*. Reporting Times 11. September 2017. https://www.ccrs.uzh.ch/en/News/pa-in-der-ccr-reporting-times-No.-11-2017.html?fontsize=big. Accessed September 25, 2018.
Aerni, P. (2017b). Principled embeddedness': How foreign direct investment may contribute to inclusive and sustainable growth in developing economies. *ATDF Journal, 9*(1/2), 3–19.
Aerni, P. (2017c). Regulatory barriers to trade: Private standards. Chapter III.27. In: T. Cottier & K. N. Schefer (Eds.), *Elgar encyclopedia of international economic law* (pp. 413–415). Edward Elgar Publishing.
Aerni, P. (2016). Coping with migration-induced urban growth: Addressing the blind spot of UN habitat. *Sustainability, 8*(8), 800. https://doi.org/10.3390/su8080800.
Aerni, P. (2015a). *The sustainable provision of environmental services: From regulation to innovation*. Heidelberg: Springer.
Aerni, P. (2015b). *Entrepreneurial rights as human rights*. Cambridge: Banson. http://pubhtml5.com/nnfz/gcoc. Accessed September 25, 2018.
Aerni, P. (2015c). Agricultural biotechnology and public attitudes: An attempt to explain the mismatch between experience and perception'. In R. Watson & V. R. Preedy (Eds.), *Genetically modified organisms in food* (pp. 149–156). Amsterdam: Elsevier Press.
Aerni, P., Nichterlein, K., Rudgard, S., & Sonnino, A. (2015). Making agricultural innovation systems (AIS) work for development in tropical countries. *Sustainability, 7*(1), 831–850. https://doi.org/10.3390/su7010831.
Aerni, P. (2013a). Resistance to agricultural biotechnology: The importance of distinguishing between weak and strong public attitudes. *Biotechnology Journal, 8*(10), 1129–1132.
Aerni, P. (2013b). *Do private standards encourage or hinder innovation?* NCCR Trade Working Paper 2013/18, University of Bern.
Aerni, P. (2011a). Connecting catch-up growth to sustainable development—A new theoretical perspective. *African Technology Development Forum (ATDF) Journal, 8*(3/4), 29–42.
Aerni, P. (2011b). Die Moralisierung der Politik als Kehrseite der Angst vor dem globalen Wandel'. In P. Aerni & K.-J Grün (Eds.), *Moral und Angst* (pp. 13–32). Göttingen: Vandenhoeck und Ruprecht Verlag.

P. Aerni, *Global Business in Local Culture*, SpringerBriefs in Economics,
https://doi.org/10.1007/978-3-030-03798-7

Aerni, P. (2011c). Interdisziplinarität an Schule und Universität: Das Fallbeispiel Geographie'. In P. Aerni & F. Oser (Eds.), *Forschung verändert Schule* (pp. 152–174). Zürich: Seismo verlag.

Aerni, P., Karapinar, B., & Häberli, C. (2011). Rethinking sustainable agriculture'. In T. Cottier & P. Delimatsis (Eds.), *The prospects of international trade regulation: From fragmentation to coherence* (pp. 169–210). Cambridge: Cambridge University Press.

Aerni, P. (2009). What is sustainable agriculture? Empirical evidence of diverging views in Switzerland and New Zealand. *Ecological Economics, 68*(6), 1872–1882.

Aerni, P. (2007a). Exploring the linkages of commerce, higher education and human development: A historical review. *ATDF Journal, 4*(2), 35–48.

Aerni, P. (2007b). Anwendungen der Agrarbiotechnologie in der Armutsbekämpfung. In U. Renz & Barbara Bleisch (Eds.), *Zu wenig: Dimensionen der Armut* (pp. 218–228). Zürich: Seismo Verlag.

Aerni, P. (2006). The principal-agent problem in international development assistance and its impact on local entrepreneurship in Africa: Time for new approaches. *ATDF Journal, 3*(2), 27–33.

Aerni, P., & Bernauer, T. (2006). Stakeholder attitudes toward GMOs in the Philippines, Mexico, and South Africa: The issue of public trust. *World Development, 34*(3), 557–575.

AHK (Delegation of German Industry and Commerce in Kenya). (2014). *Zielmarktanalyse Biogas und Biomasse Kenia. Mit Profilen der Marktakteure*. Nairobi: AHK Kenia. https://www.giz.de/fachexpertise/downloads/ahk2014-kenia-zielmarktanalyse-bioenergie.pdf. Accessed September 25, 2018.

Alarcon, P., Fèvre, E. M., Murungi, M. K., Muinde, P., Akoko, J., Dominguez-Salas, P., et al. (2017). Mapping of beef, sheep and goat food systems in Nairobi—A framework for policy making and the identification of structural vulnerabilities and deficiencies. *Agricultural systems, 152,* 1–17.

Alcácer, J., Cantwell, J., & Piscitello, L. (2016). Internationalization in the information age: A new era for places, firms, and international business networks? *Journal of International Business Studies, 47*(5), 499–512.

Allan, T., & Gordon, S. (2009). *The scalpel, the sword: The Story of Doctor Norman Bethune* (Vol. 13). Toronto: Dundurn.

Amendolagine, V., Presbitero, A., Rabellotti, R., Sanfilippo, M., & Seric, A. (2017). *FDI, global value chains, and local sourcing in developing countries*. IMF Working Paper No. 17/284. International Monetary Fund.

Andersson, U., & Forsgren, M. (2000). In search of centre of excellence: Network embeddedness and subsidiary roles in multinational corporations. *MIR: Management International Review*, 329–350.

Andersson, U., Forsgren, M., & Holm, U. (2002). The strategic impact of external networks: Subsidiary performance and competence development in the multinational corporation. *Strategic Management Journal, 23*(11), 979–996.

Appiah, A. (2006). *Cosmopolitanism: Ethics in a world of strangers*. Allen Lane.

Ariely, D. (2009). *Predictably irrational*. New York: HarperCollins.

Atkinson, R. D. & Lind, M. (2018). *Is big business really that bad?* The Atlantic, April 2018 Issue. https://www.theatlantic.com/magazine/archive/2018/04/learning-to-love-big-business/554096/. Accessed September 25, 2018.

Avins, J. (2015). *Borrowing from other cultures can be a positive exchange*. The Atlantic, October 15, 2015. https://www.theatlantic.com/politics/archive/2015/10/borrowing-from-other-cultures-can-be-a-positive-exchange/433194/. Accessed September 25, 2018.

Baker, D. (2016). *Rigged: How globalization and the rules of the modern economy were structured to make the rich richer*. Center for Economic and Policy Research.

Barnes, T. J. (2001). Retheorizing economic geography: From the quantitative revolution to the "Cultural turn". *Annals of the Association of American Geographers, 91*(3), 546–565.

Barthes. R. (2012). *Mythen des Alltags* (translation of the original French publication titled Mythologies in 1957). Suhrkamp.

Bata-Blyth, R. (2008). In Celebration of Thomas J. Bata (1914–2008). Toronto 2008.

Bang, P. F. (2016). Beyond capitalism–conceptualising ancient trade through friction, world historical context and bazaars. In J. C. M. Garcia (Ed.), *Dynamics of production in the ancient near east* (pp. 75–89). Oxbow Books.

Beckert, J. (2007). *The great transformation of embeddedness: Karl Polanyi and the new economic sociology*. MPIfG Discussion Paper 01/7. Köln: Max Planck Institut für Gesellschaftsforschung.

Berman, B. J. (1998). Ethnicity, patronage and the African state: The politics of uncivil nationalism. *African Affairs, 97*(388), 305–341.

Biggart, N. W., & Beamish, T. D. (2003). The economic sociology of conventions: Habit, custom, practice, and routine in market order. *Annual Review of Sociology, 29*, 443–464.

Bonny, S. (2017). Corporate concentration and technological change in the global seed industry. *Sustainability, 9*(9), 1632. https://doi.org/10.3390/su9091632.

Bourdieu, P. (1988). *Homo academicus*. Stanford University Press.

Braudel, F. (1982). *The wheels of commerce: Civilization and capitalism 15th–18th*, (Vol. II). London: Phoenix Press.

Braun, R. (1965). Sozialer und kultureller Wandel in einem ländlichen Industriegebiet:(Zürcher Oberland) unter Einwirkung des Maschinen-und Fabrikwesens im 19 und 20. Jahrhundert. Chronos.

Brautigam, D., & Gallagher, K. P. (2014). Bartering globalization: China's commodity-backed Financein Africa and Latin America. *Global Policy, 5*(3), 346–352.

Bugg-Levine, A., & Emerson, J. (2011). Impact investing: Transforming how we make money while making a difference. *Innovations: Technology, Governance, Globalization, 6*(3), 9–18.

Buchanan, J., & Tullock, G. (1962). *The calculus of consent*. Ann Arbor Paperbacks.

Bugra, A., & Agartan, K. (2007). *Reading Karl Polanyi for the twenty-first century: market economy as a political project*. Berlin: Springer.

Carroll, J. F. M. (2017). *Anthropology's interrogation of philosophy from the eighteenth to the twentieth century*. Lexington Books.

Chen, H., & Soltes, E. (2018). Why compliance programs fail and how to fix them. *Harvard Business Review, 96*(2), 115–125.

Ciaian, P., Pokrivcak, J., & Szegenyova, K. (2012). *Do agricultural subsidies crowd out or stimulate rural credit market institutions?* The case of EU Common Agricultural Policy.

Clark, P., & Hussey, I. (2016). Fair trade certification as oversight: An analysis of fair trade international and the small producers' symbol. *New Political Economy, 21*(2), 220–237.

Coase, R. (1937). The nature of the firm. *Economica, 4*(16), 386–405.

Committee on Sustainable Agriculture (COSA). (2014). *The COSA measuring sustainability report*. COSA. https://thecosa.org/wp-content/uploads/2014/01/The-COSA-Measuring-Sustainability-Report.pdf. Accessed September 25 2018.

Cope, J., Jack, S., & Rose, M. B. (2007). Social capital and entrepreneurship: An introduction. *International Small Business Journal, 25*(3), 213–219.

Cornell, A., Møller, J., & Skaaning, S. (2017). The real lessons of the interwar years. *Journal of Democracy, 28*(4).

Cowen, T. (2017). *The complacent class*. New York: St. Martin's Press.

Cronon, W. (1992). A place for stories: Nature, history, and narrative. *The Journal of American History, 78*(4), 1347–1376.

Current, C. B., & Tillotson, E. (2018). Hipster racism and sexism in charity date auctions: Individualism, privilege blindness and irony in the academy. *Gender and Education, 30*(4), 467–476.

Dhair, A. L. (2018). *The global leader in pay-as-you-go solar power is downsizing to stay profitable*. Quartz Africa, March 15, 2018. https://qz.com/1229170/m-kopa-solar-lays-off-kenya-uganda-staff-as-it-restructures/. Accessed September 25, 2018.

Deaton, A. (2015). *The great escape*. Princeton, NJ: Princeton University Press.

Deaton, A. (2004). *Health in an age of globalization* (No. w10669). National Bureau of Economic Research.

De Bakker, F. G., Den Hond, F., King, B., & Weber, K. (2013). Social movements, civil society and corporations: Taking stock and looking ahead. *Organization Studies, 34*(5–6), 573–593.

Deephouse, D. L., Bundy, J., Tost, L. P., & Suchman, M. C. (2017). Organizational legitimacy: Six key questions. In *The SAGE handbook of organizational institutionalism* (pp. 27–54).

Desai, M. (2003). Public goods: A historical perspective. In I. Kaul, et al. (Eds.), *Providing global public goods: Managing globalization* (pp. 63–77). New York: Oxford University Press.

Desta, M. (2002). *The law of international trade in agricultural products: From GATT 1947 to the WTO Agreement on Agriculture*. Kluwer Law International.

De Sardan, J. O. (1999). A moral economy of corruption in Africa? *The Journal of Modern African Studies, 37*(1), 25–52.

De Soto, H. (2011). *The real Mohamed Bouazizi, foreign policy*, December issue. http://foreignpolicy.com/2011/12/16/the-real-mohamed-bouazizi/.

Dewey, J. (1905). *How we think*. Boston, D.C: Heath.

Diamond, J., & Robinson, J. A. (Eds.). (2010). *Natural experiments of history*. Cambridge: Harvard University Press.

Diamond, J. (2014). Reversals of national fortune, and social science methodologies. *Proceedings of the National Academy of Sciences, 111*(50), 17709–17714.

Dorobantu, S., & Fleming, D. (2017). It's never been more important for big companies to listen to local communities. *Harvard Business Review*, November 10, 2017. https://hbr.org/2017/11/its-never-been-more-important-for-big-companies-to-listen-to-local-communities, September 25, 2018.

Dragusanu, R., Giovannucci, D., & Nunn, N. (2014). The economics of fair trade. *The Journal of Economic Perspectives, 28*(3), 217–236.

Dunning, T. (2004). Conditioning the effects of aid: Cold war politics, donor credibility, and democracy in Africa. *International Organization, 58*(2), 409–423.

Eagleton, T. (2016). *Culture*. Yale University Press.

Easterly, W. (2002). *The cartel of good intentions: Bureaucracy versus markets in foreign aid*. Center for Global Development Working Paper No. 4.

Easterly, W. (2007). *The white man's burden: Why the West's efforts to aid the rest have done so much ill and so little good*. Penguin Books.

Ecclestone, K. (2017). Behaviour change policy agendas for 'vulnerable'subjectivities: The dangers oftherapeutic governance and its new entrepreneurs. *Journal of Education Policy, 32*(1), 48–62.

Eggen, O., & Roland, K. (2013). *Western aid at a crossroads: The end of paternalism*. Palgrave Macmillan.

Eggenberger, K. (2018). When is blacklisting effective? Stigma, sanctions and legitimacy: The reputational and financial costs of being blacklisted. *Review of International Political Economy*. https://doi.org/10.1080/09692290.2018.1469529.

Ellul, J. (1990). *The Technological Bluff*. W. B. Eerdmans Publishing Co.

Etzkowitz, H. (2004). The evolution of the entrepreneurial university. *International Journal of Technology and Globalisation, 1*(1), 64–77.

Etzkowitz, H., & Zhou, C. (2017). *The triple helix: University–industry–government innovation and entrepreneurship*. Routledge.

Fagan, B. (2001). *The little ice age: How climate made history*. New York: Basic Books.

Farla, K., De Crombrugghe, D., & Verspagen, B. (2016). Institutions, foreign direct investment, and domestic investment: Crowding out or crowding in? *World Development, 88*, 1–9.

Fontaine, O., Garner, P., & Bhan, M. K. (2007). Oral rehydration therapy: The simple solution for saving lives. *British Medical Journal, 14*.

Food and Agriculture Organisation (FAO). (2013). *Livestock and biogas: Positive externalities generated by the Keekonyokie Slaughterhouse in Kiserian, Kenya.* FAO Factsheet, FAO . http://www.fao.org/fileadmin/user_upload/pes-project/docs/FAO_RPE-PES_Kiserian_Kenya.pdf. Accessed September 25, 2018.

Frei, U., & Schluep, I. (2017). From philanthropy to business: When transformation is the only way to stay true to your mission. *ATDF Journal, 9*(1), 48–55.

Freidberg, S. (2009). *Fresh: A perishable history*. Cambridge: Harvard University Press.

Freidberg, S. (2007). Supermarkets and imperial knowledge. *Cultural Geographies, 14*(3), 321–342.

Furedi, F. (2003). *Therapy culture: Cultivating vulnerability in an age of uncertainty*. United Kingdom: Psychology Press.

Furedi, F. (2016). *What's happened to the university? A sociological exploration of its infantilisation*. United Kingdom: Routledge.

Gambetta, D. (1996). *The Sicilian Mafia: The business of private protection*. United Kingdom: Harvard University Press.

Gemici, K. (2008). Karl Polanyi and the antinomies of embeddedness. *Socio-Economic Review, 6* (1), 5–33.

Graeber, D. (2011). *Debt: The first 5000 years*. New York: Melville House Publishing.

Granovetter, M. (1985). Economic action and social structure: The problem of embeddedness. *American Journal of Sociology, 91*(3), 481–510.

Granovetter, M. (1995). *Getting a job: A study of contacts and careers*. Chicago: University of Chicago Press.

Granovetter, M., & Swedberg, R. (Eds.). (2001). *The sociology of economic life* (Vol. 3). Boulder, CO: Westview Press.

Greene, J. (2014). *Moral tribes: Emotion, reason, and the gap between us and them*. Penguin.

Gilbert, N. (2012). African agriculture: Dirt poor. *Nature, 483*, 525–527.

Gintis, H. & Bowles, S. (2005). *Moral sentiments and material interests: The foundations of cooperation in economic life*. MIT press.

Gruère, G. P., & Sun, Y. (2012). *Measuring the contribution of Bt cotton adoption to India's cotton yields leap*. International Food Policy Research Institute (IFPRI) Discussion Paper, 1170.

Gruère, G. P., Mehta-Bhatt, P., & Sengupta, D. (2008). *Bt cotton and farmer suicides in India: Reviewing the evidence*. International Food Policy Research Institute (IFPRI).

Haas, P. M. (1992). Introduction: Epistemic communities and international policy coordination. *International Organization, 46*(1), 1–35.

Haidt, J. (2012). *The righteous mind. Why good people are divided by politics and religion*. New York: Vintage Press.

Haight, C. (2011). *The problem with fair trade coffee. Stanford review on social innovation*. Summer 2011. https://ssir.org/articles/entry/the_problem_with_fair_trade_coffee. Accessed September 25, 2018.

Hallaert, J.-J. (2017). Aid for trade initiative (Chapter IV.11). In: T. Cottier & K. N. Schefer (Eds.), *Elgar encyclopedia of international economic law* (pp. 603–606). United Kingdom: Edward Elgar Publishing.

Hamblin, J. D. (2013). *Arming mother nature: The birth of catastrophic environmentalism*. Oxford: Oxford University Press.

Harring, N. (2018). Trust and state intervention: Results from a Swedish survey on environmental policy support. *Environmental Science and Policy, 82*, 1–8.

Hartigan, J. (2005). *Odd tribes: Toward a cultural analysis of white people*. Duke: University Press.

Hayek, F. A. (1994, first published in 1944). *Road to Serfdom*. Chicago: University of Chicago Press.

Hays, J. C. (2009). *Globalization and the new politics of embedded liberalism*. Oxford: Oxford University Press.

Henderson, D. R. (2008). Fair trade is counterproductive–and unfair. *Economic Affairs, 28*(3), 62–64.

Henrich, J. (2006). Cooperation, punishment, and the evolution of human institutions. *Science, 311* (5769), 60–61.

Henrich, J. (2011). Angst und Moral im Tauschhandel (Interview). In P. Aerni, & K.-J. Grün (Eds.), *Moral und Angst* (pp. 45–56). Vandenhoeck and Ruprecht Verlag.

Henrich, N., & Henrich, J. P. (2007). *Why humans cooperate: A cultural and evolutionary explanation*. Oxford: Oxford University Press.

Herring, R. J. (2010). Epistemic brokerage in the bio-property narrative: Contributions to explaining opposition to transgenic technologies in agriculture. *New Biotechnology, 27*(5), 614–622.

Hidalgo, C. (2015). Why information grows: *The evolution of Order, from Atoms to Economies*. Basic Books.

Himick, D., & Brivot, M. (2018). Carriers of ideas in accounting standard-setting and financialization: The role of epistemic communities. *Accounting, Organizations and Society, 66,* 29–44.

Hindmoor, A. (1998). The importance of being trusted: Transaction costs and policy network theory. *Public Administration, 76*(1), 25–43.

Hirschman, A. O. (1992). *Rival views of market society and other recent essays* (Originally published by Viking Penguin in 1986). Cambridge: Harvard University Press.

Hirschman, A. O. (1970). *Exit, voice, and loyalty: responses to decline in firms, organizations, and states*. United Kingdom: Harvard University Press.

Holenstein, A., Maissen, T., & Prak, M. R. (Eds.). (2008). *The republican alternative: The Netherlands and Switzerland compared.* Amsterdam: Amsterdam University Press.

Holenstein, A., Kury, P. & Schulz, C. (2018). *Migrationsgeschichte: Von den Anfängen bis zur Gegenwart.* Hier und Jetzt Verlag.

Holmes, C. (2013). Ignorance, denial, internalisation, and transcendence: A post-structural perspective on Polanyi's double movement. *Review of International Studies, 39*(2), 273–290.

Hopkin, J. (2017). When Polanyi met Farage: Market fundamentalism, economic nationalism, and Britain's exit from the European Union. *The British Journal of Politics and International Relations, 19*(3), 465–478.

Huybrechts, B., Nicholls, A., & Edinger, K. (2017). Sacred alliance or pact with the devil? How and why social enterprises collaborate with mainstream businesses in the fair trade sector. *Entrepreneurship and Regional Development, 29*(7–8), 586–608.

Ice, R. (1991). Corporate publics and rhetorical strategies: The case of Union Carbide's Bhopal crisis. *Management Communication Quarterly, 4*(3), 341–362.

Jacobs, J. (1970). *The economy of cities.* Vintage Books.

Jones, C. I., & Romer, P. M. (2010). The new Kaldor facts: Ideas, institutions, population, and human capital. *American Economic Journal: Macroeconomics, 2*(1), 224–245.

Juma, C. (2011). *New harvest: Agricultural innovation in Africa.* Oxford: Oxford University Press.

Kaplan, R. S., Serafeim, G., & Tugendhat, E. (2018). Inclusive growth: Profitable strategies for tackling poverty and inequality. *Harvard Business Review, 96*(1), 127–133.

Karuga, J. (2015). *Turning blood to power, Maasai pastoralists begin bottling biogas*. Reuters Environment. https://www.reuters.com/article/us-kenya-biogas/turning-blood-to-power-maasai-pastoralists-begin-bottling-biogas-idUSKBN0KK06U20150111. Accessed September 25, 2018.

Keller, E. (2015). *Beyond the lens of conservation: Malagasy and Swiss imaginations of one another* (Vol. 20). Berghahn Books.

Khan, A. (2007). *Growth, employment and poverty: An analysis of the vital nexus based on some recent UNDP and ILO/SIDA studies*. United Nations, Department of Economics and Social Affairs.

Köckritz, A. (2017). Näh schnell, näh sauber. Die Zeit, 13.12.2017. https://www.zeit.de/2017/52/aethopien-textilindustrie-hm-modekonzern-frauen-produktion. Accessed September 25, 2018.

Kojève, A. (1975). Hegel. Herausgegeben von Iring Fetscher. Erweiterte Ausgabe.

Kock, M. A., & Gould, C. (2013). Adapting IP to an evolving agricultural innovation landscape. *WIPO Magazine*, 01028.

Kock. (2018). *Is there a need to evolve IP: An open innovation enablement (IE) model for technologies addressing essential human needs*. Presentation at the University of Zurich on April 11, 2018.

Kolins Givan, R., Roberts, K. M, & Soule S. A. (2010). *The diffusion of social movements: Actors, mechanisms, and political effects*. Cambridge: Cambridge University Press.

Kolk, A., Kourula, A. E., & Pisani, N. (2017). *Multinational enterprises and the sustainable development goals: What do we know and how to proceed?* Forthcoming in Transnational Corporations.

König, T., & Gorman. M. E. (2017). The challenge of funding interdisciplinary research: A look inside policy research funding agencies. In R. Frodeman, J. T. Klein, & R. C. D. S. Pacheco (Eds.), *The oxford handbook of interdisciplinarity*. Oxford: Oxford University Press.

König, T. (2018). Political science and the European research council. *European Political Science*, 1–19.

Kovács, L., Csupor, D., Lente, G., & Gunda, T. (2014). *100 chemical myths : Misconceptions, misunderstandings, explanations*. Berlin: Springer.

Krishna, V. V., & Qaim, M. (2012). Bt cotton and sustainability of pesticide reductions in India. *Agricultural Systems, 107*, 47–55.

Kuttner, R. (2017). The man from Red Vienna. *The New York Review of Books*. December 2017 Issue, 55–57.

Lässig, S. (2016). The history of knowledge and the expansion of the historical research agenda. *Bulletin of the German Historical Bulletin, 58*, 29–58.

Lander, C. (2008). *Stuff white people like: A definitive guide to the unique taste of millions*. Random House.

Lanz, R., & Miroudot, S. (2011). *Intra-firm trade: Patterns, determinants and policy implications*. OECD Trade Policy Papers, No. 114. Paris: OECD Publishing.

Laufer, P. (2014). *Organic: A journalist's quest to discover the truth behind food labeling*. Lyons Press.

Lewis, A. (1955, 2003). *The theory of economic growth*. Routledge.

Lott, J. (2015). *Natural deception: A sobering look at the truth behind the organic food industry*. Create Space publishing.

Lu, H., Schmidpeter, R., Capaldi, N., & Zu, L. (Eds.). (2017). *Building new bridges between business and society: Recent research and new cases in CSR, sustainability, ethics and governance*. Springer.

Luhmann, N. (1993). *Communication and social order: Risk: A sociological theory*. Transaction Publishers.

Luo, Y., & Zhang, H. (2016). Emerging market MNEs: Qualitative review and theoretical directions. *Journal of International Management, 22*(4), 333–350.

Makita, R., & Tsuruta, T. (2017). *Fair trade and organic initiatives in asian agriculture: The hidden realities*. Routledge.

Mandela, N. (1994, 2013). *Long walk to freedom*, UK: Hachette.

Martin, E. (2012). On one forgotten cause of the Arab spring: The lack of economic freedom. *Economic Affairs, 32*(3), 94–96.

Maris, V., Huneman, P., Coreau, A., Kéfi, S., Pradel, R., & Devictor, V. (2018). Prediction in ecology: Promises, obstacles and clarifications. *Oikos, 127*(2), 171–183.

Martinez-Alier, J. (2012). Environmental justice and economic degrowth: An alliance between two movements. *Capitalism Nature Socialism, 23*(1), 51–73.

Matthews, D., & Zech, H. (Eds.). (2017). *Research handbook on intellectual property and the life sciences*. United Kingdom: Edward Elgar Publishing.
McCormick, T. J. (1995). *America's half-century: United States foreign policy in the cold war and after*. JHU Press.
Merry, S. E., & Conley, J. M. (2011). Measuring the world: Indicators, human rights, and global governance. *Current Anthropology, 52*(S3), 000.
Meyer, K. E., Mudambi, R., & Narula, R. (2011). Multinational enterprises and local contexts: The opportunities and challenges of multiple embeddedness. *Journal of Management Studies, 48* (2), 235–252.
Miller, D. (2012). *Consumption and its consequences*. Malden MA: Polity Press.
Molina-Morales, F. X., & Martínez-Fernández, M. T. (2010). Social networks: Effects of social capital on firm innovation. *Journal of Small Business Management, 48*(2), 258–279.
Morris, S. (2000). Defining the nonprofit sector: Some lessons from history. *Voluntas: International Journal of Voluntary and Nonprofit Organizations, 11*(1), 25–43.
Naam, R. (2013). *The infinite resource: The power of ideas on a finite planet*. United States: University of England Press.
Nathans, B. (2017). *Bolshevism's new believers. The New York Review of Books*, November 23, 2017. http://www.nybooks.com/articles/2017/11/23/bolshevisms-new-believers/. Accessed September 25, 2018.
Nelkin, D. (1995) *Selling science. How the press covers science and technology*. W.W. Freeman & Co.
New Zimbabwe. (2015). *Sure footed Bata begins long trek toward former glory*, New Zimbabwe, August 18, 2015. https://www.newzimbabwe.com/sure-footed-bata-begins-long-trek-towards-former-glory/. Accessed September 25, 2018.
Nyende, P. (2011). *Building networks for market access*. Rome: FAO.
Nichols, T. (2017) The death of expertise: The campaign against established knowledge and why it matters. Oxford: Oxford University Press.
Nickerson, R. S. (1998). Confirmation bias: A ubiquitous phenomenon in many guises. *Review of General Psychology, 2*(2), 175–220.
OECD. (2002). *Foreign direct investment for development: Maximising benefits, minimizing costs*. OECD Publishing.
Olpadwala, P. (2012). The 'Enigma' of hunger: Comment. *Social Scientist, 40*(9/10), 91–108. http://www.jstor.org/stable/23338889. Accessed September 25, 2018.
Olson, M. (2008 first published in1982). *The rise and decline of nations: Economic growth, stagflation, and social rigidities*. Yale University Press.
Olson, M. (2009 first published in 1965). *The logic of collective action* (Vol. 124). United Kingdom: Harvard University Press.
Paarlberg, R., & Paarlberg, R. L. (2013). *Food politics: What everyone needs to know*. Oxford: Oxford University Press.
Panitch, L. (2015). Capital and politics. *Perspectives on Politics, 13*(4), 1075–1083.
Pepinsky, T. B. (2017). *Dis-embedding liberal internationalism*. Available at SSRN: https://ssrn.com/abstract=3054001. Accessed September 25, 2018.
Petley, J. (Ed.). (2013). *Media and public shaming: Drawing the boundaries of disclosure*. IB Tauris.
Ping, W. (2011). *A brief history of standards and standardization organisation: A chinese perspective*. East-West Center Working Papers. Economic Series No 117. http://www.eastwestcenter.org/sites/default/files/private/econwp117.pdf. Accessed September 25, 2018.
Polanyi, K. (1978, first published in 1944). The great transformation: Politische und ökonomische Ursprünge von Gesellschaften und Wirtschaftssystemen (Übers. von Heinrich Jelinek). Suhrkamp, Frankfurt/Main.
Polanyi, K. (2001, first published in 1944). *The great zransformation: The political and economic origins of our times*. With a foreword by Joseph Stiglitz. Beacon Press.

Porou, T. (2013). Enduring land and cultural sustainability for Ngāti Porou through afforestation subsidies. In *Factsheet prepared for the multi-stakeholder dialogue on investing in environmental services*, September 12–13, 2013. Rome: FAO.

Power, M. (1997). *The global audit society*. Oxford: Oxford University Press.

Ramone, J. (2013). Tastes of paradise: The 'Fair' trade myth. In P. Bennett & J. McDougall (Eds.), *Barthes' mythologies today* (pp. 24–26). Routledge.

Rangan, H. (2000). *Of myths and movements. Rewriting Chipko into Himalayan History*. Verso.

Rangan, H., & Garb, Y. (1996). *Conflicting histories: Negotiating reality and myth around the Chipko movement*. Mimeo.

Reckwitz, A. (2018). Das Band zerreist. Die Mittelschicht driftet auseinander. Feuilleton, die Zeit, February 22, 2018, p. 43.

Ren, S., Wood, R., & Zhu, Y. (2015). *Business leadership development in China*. Routledge.

Ridley, D. B. (2005). Price differentiation and transparency in the global pharmaceutical marketplace. *Pharmacoeconomics, 23*(7), 651–658.

Robinson, W. I. (2014). *Global capitalism and the crisis of humanity*. Cambridge: Cambridge University Press.

Robin, C., & Wyss Bisang, B. (2017). Fair Trade & Co: Die Pionierphase ist vorbei. Die Volkswirtschaft 7/2017, 29–31.

Rodrik, D. (2011). *The globalization paradox*. W. W. Norton and Company.

Rodrik, D. (2015). *Economics rules: The rights and wrongs of the dismal science*. W. W. Norton and Company.

Rodrik, D. (2018). *In defense of economic populism*. Project Syndicate, 9. January 2018. https://www.project-syndicate.org/commentary/defense-of-economic-populism-by-dani-rodrik-2018-01?barrier=accesspaylog. Accessed September 25, 2018.

Ronald, P. C., & Adamchak, R. W. (2008). *Tomorrow's table: Organic farming, genetics, and the future of food*. Oxford: Oxford University Press.

Romer, P. (1994). New goods, old theory, and the welfare costs of trade restrictions. *Journal of development Economics, 43*(1), 5–38.

Romer, P. M. (1996). Preferences, promises, and the politics of entitlement. In *Individual and social responsibility: Child care, education, medical care, and long-term care in America* (pp. 195–228). Chicago: University of Chicago Press.

Romer, P. M. (2010). What parts of globalization matter for catch-up growth? *American Economic Review, 100*(2), 94–98.

Ruggie, J. G. (1982). International regimes, transactions, and change: Embedded liberalism in the postwar economic Order. *International Organisation, 36*(2), 379–415.

Ruggie, J. G. (2008). *Embedding global markets: An enduring challenge*. Ashgate.

Rogers, J. R. (2017). *Karl Polanyi's irrelevance to today's policy debates. 'Law and Liberty'*, September 18, 2017. https://www.lawliberty.org/2017/09/18/karl-polanyis-irrelevance-to-todays-policy-debates/. Accessed September 25, 2018.

Rowe, G., & Wright, G. (2001). Differences in expert and lay judgments of risk: myth or reality? *Risk Analysis, 21*(2), 341–356.

Sahlins, M. (2013). *Stone-age economics*. London: Routledge Classics.

Sandefur, T. (2010). *The right to earn a living*. Washington, DC: Cato Institute.

Sauerländer, D. (2016). Die Schweiz als Auswanderungsland. *Swiss Archives of Neurology, Psychiatry and Psychotherapy, 167*(7), 208–214.

Scafidi, S. (2005) *Who owns culture? Appropriation and authencity in American law*. Rutgers University Press.

Scherer, F. M., & Watal, J. (2002). Post-TRIPS options for access to patented medicines in developing nations. *Journal of International Economic Law, 5*(4), 913–939.

Scherer, A. G., Palazzo, G., & Baumann, D. (2006). Global rules and private actors: Toward a new role of the transnational corporation in global governance. *Business Ethics Quarterly, 16*(4), 505–532.

Scheyvens, R., Banks, G., & Hughes, E. (2016). The private sector and the SDGs: The need to move beyond ‘business as usual’. *Sustainable Development, 24*(6), 371–382.

Schluep, I. (2017). Embeddedness»: die Einbettung von Firmen als Voraussetzung für Nachhaltigkeit. *Austrian Paper, 16,* 1–10.

Schluep, I., & Aerni, P. (2016). *When corporatism leads to corporate governance failure: The case of the Swiss watch industry*. Cambridge Banson. Available online: http://www.ourplanet.com/swisswatches/index.php.

Schuessler, E., Frenkel, S. J., & Wright, C. F. (2018). Governance of labor standards in Australian and German garment supply chains: The impact of Rana Plaza. *ILR Review*. https://doi.org/10.1177/0019793918771360.

Schumpeter, J. A. (2008, first published in 1942). *Capitalism, socialism and democracy*. Harper Collins Publishers.

Schwab, F. (2017). Krieg der Öko-Bananen. Weltwoche Nr 38, 39.

Scott, A. J. (2017). A perspective of economic geography. In *Economy* (pp. 3–23). Routledge.

Scott-Kennel, J., & Michailova, S. (2016). Subsidiary internal and external embeddedness: Trade-up and trade-off. *International Journal of Multinational Corporation Strategy, 1*(2), 133–154.

Sennett, R. (1977). *The fall of public man*. W.W. Norton and Company.

Shaikh, A. (2016). *Capitalism: Competition, conflict, crises*. Oxford: Oxford University Press.

Shapiro, S. P. (1987). The social control of impersonal trust. *American Journal of Sociology, 93* (3), 623–658.

Sheller, M. (2017). From spatial turn to mobilities turn. *Current Sociology, 65*(4), 623–639.

Shiva, V. (2016). *Water wars: Privatization, pollution, and profit*. North Atlantic Books.

Slezkine, Y (2017). *The house of government: A saga of the Russian revolution*. New Jersey: Princeton University Press.

Slovic, P., & Lichtenstein, S. (2006). *The construction of preference*. Cambridge: Cambridge University Press.

Smart, R. D., Blum, M., & Wesseler, J. (2017). Trends in approval times for genetically engineered crops in the United States and the European Union. *Journal of Agricultural Economics, 68*(1), 182–198.

Smith, V. L. (2003). Constructivist and ecological rationality in economics. *American Economic Review, 93*(3), 465–508.

Southgate, D., & Roberts, L. (2016). *Globalized fruit, local entrepreneurs*. United States: University of Pennsylvania Press.

Specter, M. (2014). *Seeds of doubt: An activist's controversial crusade against genetically modified crops*. The New Yorker, August 25, 2014 Issue.

Stehr, N. (2008). *Moral markets: How knowledge and affluence change consumers and products*. Routledge.

Sunstein, C. (2007). *Republic.com 2.0*. New Jersey: Princeton University Press.

Suri, T., & Jack, W. (2016). The long-run poverty and gender impacts of mobile money. *Science, 354*(6317), 1288–1292.

Tauger, M. B. (2011). *Agriculture in world history*. Routledge.

The Economicst. (2012). Going bananas: Chiquita has tried hard to be good—But not getting credit for it. *The Economist*. March 31, 2012 (available online: https://www.economist.com/node/21551500).

Tugendhat, E. (2017). *Transforming the Ugandan maize system: A positive impact case study*. Washington, DC: Palladium.

Tyler, C (2017). *The Complacent Class: The Self-defeating Quest for the American Dream*. St. Martin's Press.

Uekoetter, F (2006). *The green and the brown: A history of conservation in Nazi Germany*. Cambridge: Cambridge University Press.

United Nations Conference on Trade and Development (UNCTAD). (2013). World Investment Report 2013: Investment and trade for development. UNCTAD Publication UNCTAD/WIR/2013. UNCTAD Geneva. http://unctad.org/en/pages/PublicationWebflyer.aspx?publicationid=588. Accessed September 25, 2018.

United Nations Conference on Trade and Development (UNCTAD). (2017a). World Investment Report 2017: Investment and the Digital Economy. UNCTAD Publication UNCTAD/WIR/2017. UNCTAD Geneva. http://unctad.org/en/pages/PublicationWebflyer.aspx?publicationid=1782. Accessed September 25, 2018.

United Nations Conference on Trade and Development (UNCTAD). (2017b). New innovation approaches to support the implementation of the sustainable development goals. UNCTAD Publication UNCTAD/DTL/STICT/2017/4. UNCTAD Geneva. http://unctad.org/en/pages/PublicationWebflyer.aspx?publicationid=1775. Accessed September 25, 2018.

United Nations Conference on Trade and Development (UNCTAD). (2017c). The role of science, technology and innovation in ensuring food security by 2030. UNCTAD Publication UNCTAD/DTL/STICT/2017/5. UNCTAD Geneva. http://unctad.org/en/pages/Publication Webflyer.aspx?publicationid=1774. Accessed September 25, 2018.

United Nations Development Program (UNDP). (2015). Report on China's implementation of the millennium development goals (2000–2015). New York: UNPD. http://www.cn.undp.org/content/china/en/home/library/mdg/mdgs-report-2015-.html. Accessed September 25, 2018.

United Nations Development Program (UNDP). (2017). *Spark, scale and sustain: Innovation for the sustainable development goals*. New York: UNDP Innovation Facility Report. http://www.undp.org/content/undp/en/home/librarypage/development-impact/spark-_-scale-_-sustain—2016-year-in-review.html. Accessed September 25, 2018.

Urech, F. (2018). Weltweit findet heute jede zweite mobile Geldtransaktion in Kenya statt. NZZ, January 16, 2018 . https://www.nzz.ch/international/der-weltmarktfuehrer-aus-afrika-ld.1338079. Accessed September 25, 2018.

Uzzi, B. (1996). The sources and consequences of embeddedness for the economic performance of organizations: The network effect. *American sociological review*, 674–698.

Van Winden, F. (2007). Affective public choice. In P. Schwartz (Ed.), *Public choice and the challenges of democracy* (pp. 45–61). Edward Elgar Publishing.

Walter, J., & Schofield, R. (1991). *Famine, disease and social order in early modern society*. Cambridge: Cambridge University Press.

Warsh. D. (2006). *Knowledge and the wealth of nations. A story of economic discovery*. W. W. Norton & Company.

Watal, J. (2014). Is TRIPS a balanced agreement from the perspective of recent free trade agreements?. In J. Drexl, H. G. Ruse-Khan, & S. Nadde-Phlix (Eds.), *EU bilateral trade agreements and intellectual property: For better or worse*? (pp. 41–57). Berlin, Heidelberg: Springer.

Watkins, A., Papaioannou, T., Mugwagwa, J., & Kale, D. (2015). National innovation systems and the intermediary role of industry associations in building institutional capacities for innovation in developing countries: A critical review of the literature. *Research Policy, 44*(8), 1407–1418.

Weber, M. (1992). Soziologie, Universalgeschichtliche Analysen. In Politik (Ed.) *Johannes Wincklemann*. Alfred Kröner Verlag.

Wettstein, F. (2010). The duty to protect: Corporate complicity, political responsibility, and human rights advocacy. *Journal of Business Ethics, 96*(1), 33–47.

Williams, J. (2016). *Academic freedom in an age of conformity*. Palgrave Macmillan.

World Bank. (2017). *Global economic prospects: Arm's-length trade: A source of post-crisis trade weakness*. Special Focus 2, June 2017. Washington, DC: World Bank.

World Trade Organisation (WTO). (2016) *World trade report: Levelling the trading field for SMEs*. Geneva: WTO. https://www.wto.org/english/res_e/publications_e/wtr16_e.htm. Accessed September 25, 2018.

Woetzel, J. R. (1989). *China's economic opening to the outside world: The politics of empowerment*. Praeger Publishers.

Yingming, Z. (2009). *Analysis of industrial clusters in China*. Boca Raton: CRC Press.

Zafirovski, M. (2002). *A primer on economic sociology: The duality of structure in markets*. Nova Science Publisers Inc.

Zeitz, J. (2017). The real history of American immigration. *Politico Magazine*, August 6, 2017 https://www.politico.com/magazine/story/2017/08/06/trump-history-of-american-immigration-215464. Accessed September 25, 2018.

Zubok, V. M. (2009). *A failed empire: The Soviet Union in the Cold War from Stalin to Gorbachev*. University of North Carolina Press.

MIX
Papier aus verantwortungsvollen Quellen
Paper from responsible sources
FSC® C105338
FSC
www.fsc.org

If you have any concerns about our products,
you can contact us on
ProductSafety@springernature.com

In case Publisher is established outside the EU,
the EU authorized representative is:
Springer Nature Customer Service Center GmbH
Europaplatz 3, 69115 Heidelberg, Germany

Printed by Libri Plureos GmbH
in Hamburg, Germany